HARCOURT
Science

Harcourt School Publishers

Orlando • Boston • Dallas • Chicago • San Diego

www.harcourtschool.com

Cover Image
This butterfly is a Red Cracker. It is almost completely red on its underside. It is called a cracker because the males make a crackling sound as they fly. The Red Cracker is found in Central and South America.

Copyright © 2000 by Harcourt, Inc.

All rights reserved. No part of this publication may be reproduced or transmitted in any form or by any means, electronic or mechanical, including photocopy, recording, or any information storage and retrieval system, without permission in writing from the publisher.

Requests for permission to make copies of any part of the work should be mailed to the following address:

School Permissions, Harcourt, Inc.
6277 Sea Harbor Drive
Orlando, FL 32887-6777

HARCOURT and the Harcourt Logo are trademarks of Harcourt, Inc.

sciLINKS is owned and provided by the National Science Teachers Association. All rights reserved.

Smithsonian Institution Internet Connections owned and provided by the Smithsonian Institution. All other material owned and provided by Harcourt School Publishers under copyright appearing above.

The name of the Smithsonian Institution and the Sunburst logo are registered trademarks of the Smithsonian Institution. The copyright in the Smithsonian website and Smithsonian website pages are owned by the Smithsonian Institution.

Printed in the United States of America

ISBN 0-15-315686-4   UNIT A
ISBN 0-15-315687-2   UNIT B
ISBN 0-15-315688-0   UNIT C
ISBN 0-15-315689-9   UNIT D
ISBN 0-15-315690-2   UNIT E
ISBN 0-15-315691-0   UNIT F

Authors

Marjorie Slavick Frank
Former Adjunct Faculty Member at Hunter, Brooklyn, and Manhattan Colleges
New York, New York

Robert M. Jones
Professor of Education
University of Houston-Clear Lake
Houston, Texas

Gerald H. Krockover
Professor of Earth and Atmospheric Science Education
School Mathematics and Science Center
Purdue University
West Lafayette, Indiana

Mozell P. Lang
Science Education Consultant
Michigan Department of Education
Lansing, Michigan

Joyce C. McLeod
Visiting Professor
Rollins College
Winter Park, Florida

Carol J. Valenta
Vice President—Education, Exhibits, and Programs
St. Louis Science Center
St. Louis, Missouri

Barry A. Van Deman
Science Program Director
Arlington, Virginia

Senior Editorial Advisor

Napoleon Adebola Bryant, Jr.
Professor Emeritus of Education
Xavier University
Cincinnati, Ohio

Program Advisors

Michael J. Bell
Assistant Professor of Early Childhood Education
School of Education
University of Houston-Clear Lake
Houston, Texas

George W. Bright
Professor of Mathematics Education
The University of North Carolina at Greensboro
Greensboro, North Carolina

Pansy Cowder
Science Specialist
Tampa, Florida

Nancy Dobbs
Science Specialist, Heflin Elementary
Alief ISD
Houston, Texas

Robert H. Fronk
Head, Science/Mathematics Education Department
Florida Institute of Technology
Melbourne, Florida

Gloria R. Guerrero
Education Consultant
Specialist in English as a Second Language
San Antonio, Texas

Bernard A. Harris, Jr.
Physician and Former Astronaut
(STS 55—*Space Shuttle Columbia*, STS 63—*Space Shuttle Discovery*)
Vice President, SPACEHAB Inc.
Houston, Texas

Lois Harrison-Jones
Education and Management Consultant
Dallas, Texas

Linda Levine
Educational Consultant
Orlando, Florida

Bertie Lopez
Curriculum and Support Specialist
Ysleta ISD
El Paso, Texas

Kenneth R. Mechling
Professor of Biology and Science Education
Clarion University of Pennsylvania
Clarion, Pennsylvania

Nancy Roser
Professor of Language and Literacy Studies
University of Texas, Austin
Austin, Texas

Program Advisor and Activities Writer

Barbara ten Brink
Science Director
Round Rock Independent School District
Round Rock, Texas

Reviewers and Contributors

Dorothy J. Finnell
Curriculum Consultant
Houston, Texas

Kathy Harkness
Retired Teacher
Brea, California

Roberta W. Hudgins
Teacher, W. T. Moore Elementary
Tallahassee, Florida

Libby Laughlin
Teacher, North Hill Elementary
Burlington, Iowa

Teresa McMillan
Teacher-in-Residence
University of Houston-Clear Lake
Houston, Texas

Kari A. Miller
Teacher, Dover Elementary
Dover, Pennsylvania

Julie Robinson
Science Specialist, K-5
Ben Franklin Science Academy
Muskogee, Oklahoma

Michael F. Ryan
Educational Technology Specialist
Lake County Schools
Tavares, Florida

Judy Taylor
Teacher, Silvestri Junior High School
Las Vegas, Nevada

UNIT A

LIFE SCIENCE
Plants and Animals

Chapter 1 | How Plants Grow — A2

Lesson 1—What Do Plants Need? — A4
Lesson 2—What Do Seeds Do? — A10
Lesson 3—How Do Plants Make Food? — A18
   Science and Technology • Drought-Resistant Plants — A24
   People in Science • George Washington Carver — A26
   Activities for Home or School — A27
Chapter Review and Test Preparation — A28

Chapter 2 | Types of Animals — A30

Lesson 1—What Is an Animal? — A32
Lesson 2—What Are Mammals and Birds? — A40
Lesson 3—What Are Amphibians, Fish, and Reptiles? — A48
   Science Through Time • Discovering Animals — A58
   People in Science • Rodolfo Dirzo — A60
   Activities for Home or School — A61
Chapter Review and Test Preparation — A62

Unit Project Wrap Up — A64

UNIT B

LIFE SCIENCE
Plants and Animals Interact

Chapter 1	**Where Living Things are Found**	**B2**
Lesson 1—What Are Ecosystems?	B4
Lesson 2—What Are Forest Ecosystems?	B10
Lesson 3—What Is a Desert Ecosystem?	B18
Lesson 4—What Are Water Ecosystems?	B24
Science and Technology • Using Computers to Describe the Environment	B32
People in Science • Margaret Morse Nice	B34
Activities for Home or School	B35
Chapter Review and Test Preparation	B36

Chapter 2	**Living Things Depend on One Another**	**B38**
Lesson 1—How Do Animals Get Food?	B40
Lesson 2—What Are Food Chains?	B46
Lesson 3—What Are Food Webs?	B52
Science Through Time • People and Animals	B58
People in Science • Akira Akubo	B60
Activities for Home or School	B61
Chapter Review and Test Preparation	B62

Unit Project Wrap Up   B64

UNIT C

EARTH SCIENCE
Earth's Land

Chapter 1 — Rocks, Minerals, and Fossils — C2
- Lesson 1—What Are Minerals and Rocks? — C4
- Lesson 2—How Do Rocks Form? — C10
- Lesson 3—What Are Fossils? — C18
  - Science Through Time • Discovering Dinosaurs — C24
  - People in Science • Charles Langmuir — C26
  - Activities for Home or School — C27
- Chapter Review and Test Preparation — C28

Chapter 2 — Forces That Shape the Land — C30
- Lesson 1—What Are Landforms? — C32
- Lesson 2—What Are Slow Landform Changes? — C38
- Lesson 3—What Are Rapid Landform Changes? — C46
  - Science and Technology • Earthquake-Proof Buildings — C52
  - People in Science • Scott Rowland — C54
  - Activities for Home or School — C55
- Chapter Review and Test Preparation — C56

Chapter 3 — Soils — C58
- Lesson 1—How Do Soils Form? — C60
- Lesson 2—How Do Soils Differ? — C66
- Lesson 3—How Can People Conserve Soil? — C72
  - Science and Technology • Farming with GPS — C78
  - People in Science • Diana Wall — C80
  - Activities for Home or School — C81
- Chapter Review and Test Preparation — C82

Chapter 4 — Earth's Resources — C84
- Lesson 1— What Are Resources? — C86
- Lesson 2— What Are Different Kinds of Resources? — C92
- Lesson 3— How Can We Conserve Earth's Resources? — C98
  - Science and Technology • Recycling Plastic to Make Clothing — C106
  - People in Science • Marisa Quinones — C108
  - Activities for Home or School — C109
- Chapter Review and Test Preparation — C110

Unit Project Wrap Up — C112

UNIT D

EARTH SCIENCE
Cycles on Earth and In Space

Chapter 1 — The Water Cycle — D2
Lesson 1—Where Is Water Found on Earth? — D4
Lesson 2—What Is the Water Cycle? — D14
   Science and Technology • A Filter for Clean Water — D20
   People in Science • Lisa Rossbacher — D22
   Activities for Home or School — D23
Chapter Review and Test Preparation — D24

Chapter 2 — Observing Weather — D26
Lesson 1—What Is Weather? — D28
Lesson 2—How Are Weather Conditions Measured? — D34
Lesson 3—What Is a Weather Map? — D42
   Science and Technology • Controlling Lightning Strikes — D48
   People in Science • June Bacon-Bercey — D50
   Activities for Home or School — D51
Chapter Review and Test Preparation — D52

Chapter 3 — Earth and Its Place in the Solar System — D54
Lesson 1—What Is the Solar System? — D56
Lesson 2—What Causes Earth's Seasons? — D66
Lesson 3—How Do the Moon and Earth Interact? — D74
Lesson 4—What Is Beyond the Solar System? — D82
   Science Through Time • Sky Watchers — D90
   People in Science • Mae C. Jemison — D92
   Activities for Home or School — D93
Chapter Review and Test Preparation — D94

Unit Project Wrap Up — D96

UNIT E

PHYSICAL SCIENCE
Investigating Matter

Chapter 1 — Properties of Matter — E2
Lesson 1—What Are Physical Properties of Matter? — E4
Lesson 2—What Are Solids, Liquids, and Gases? — E14
Lesson 3—How Can Matter Be Measured? — E20
   Science Through Time • Classifying Matter — E30
   People in Science • Dorothy Crowfoot Hodgkin — E32
   Activities for Home or School — E33
Chapter Review and Test Preparation — E34

Chapter 2 — Changes in Matter — E36
Lesson 1—What Are Physical Changes? — E38
Lesson 2—What Are Chemical Changes? — E44
   Science and Technology • Plastic Bridges — E50
   People in Science • Enrico Fermi — E52
   Activities for Home or School — E53
Chapter Review and Test Preparation — E54

Unit Project Wrap Up — E56

UNIT F

PHYSICAL SCIENCE
Exploring Energy and Forces

Chapter 1 — Heat — F2
- Lesson 1—What Is Heat? — F4
- Lesson 2—How Does Thermal Energy Move? — F12
- Lesson 3—How Is Temperature Measured? — F18
  - Science and Technology • Technology Delivers Hot Pizza — F24
  - People in Science • Percy Spencer — F26
  - Activities for Home or School — F27
- Chapter Review and Test Preparation — F28

Chapter 2 — Light — F30
- Lesson 1—How Does Light Behave? — F32
- Lesson 2—How Are Light and Color Related? — F42
  - Science Through Time • Discovering Light and Optics — F48
  - People in Science • Lewis Howard Latimer — F50
  - Activities for Home or School — F51
- Chapter Review and Test Preparation — F52

Chapter 3 — Forces and Motion — F54
- Lesson 1– How Do Forces Cause Motion? — F56
- Lesson 2—What Is Work? — F64
- Lesson 3—What Are Simple Machines? — F68
  - Science and Technology • Programmable Toy Building Bricks — F74
  - People in Science • Christine Darden — F76
  - Activities for Home or School — F77
- Chapter Review and Test Preparation — F78

Unit Project Wrap Up — F80

References — R1
- Science Handbook — R2
- Health Handbook — R11
- Glossary — R46
- Index — R54

Using Science Process Skills

When scientists try to find an answer to a question or do an experiment, they use thinking tools called process skills. You use many of the process skills whenever you think, listen, read, and write. Think about how these students used process skills to help them answer questions and do experiments.

Maria is interested in birds. She carefully observes the birds she finds. Then she uses her book to identify the birds and learn more about them.

Try This Find something outdoors that you want to learn more about. Use your senses to observe it carefully.

Talk About It What senses does Maria use to observe the birds?

Process Skills

Observe — use your senses to learn about objects and events

Charles finds rocks for a rock collection. He observes the rocks he finds. He compares their colors, shapes, sizes, and textures. He classifies them into groups according to their colors.

Try This Use the skills of comparing and classifying to organize a collection of objects.

Talk About It What other ways can Charles classify the rocks in his collection?

Process Skills

Compare—identify characteristics of things or events to find out how they are alike and different

Classify—group or organize objects or events in categories based on specific characteristics

Katie measures her plants to see how they grow from day to day. Each day after she **measures** she **records the data**. Recording the data will let her work with it later. She **displays the data** in a graph.

Try This Find a shadow in your room. Measure its length each hour. Record your data, and find a way to display it.

Talk About It How does displaying your data help you communicate with others?

Process Skills

Measure — compare mass, length, or capacity of an object to a unit, such as gram, centimeter, or liter

Record Data — write down observations

Display Data — make tables, charts, or graphs

An ad about low-fat potato chips claims that low-fat chips have half the fat of regular potato chips. Tani **plans and conducts an investigation** to test the claim.

Tani labels a paper bag Regular and Low-Fat. He finds two chips of each kind that are the same size, and places them above their labels. He crushes all the chips flat against the bag. He sets the stopwatch for one hour.

Tani **predicts** that regular chips will make larger grease spots on the bag than low-fat chips. When the stopwatch signals, he checks the spots. The spots above the Regular label are larger than the spots above the Low-Fat label. Tani **infers** that the claim is correct.

Try This Plan and conduct an investigation to test claims for a product. Make a prediction, and tell what you infer from the results.

Talk About It Why did Tani test potato chips of the same size?

Process Skills

Plan and conduct investigations—identify and perform the steps necessary to find the answer to a question

Predict—form an idea of an expected outcome based on observations or experience

Infer—use logical reasoning to explain events and make conclusions

You will have many opportunities to practice and apply these and other process skills in *Harcourt Science*. An exciting year of science discoveries lies ahead!

Safety in Science

Here are some safety rules to follow.

1) Think ahead. Study the steps and safety symbols of the investigation so you know what to expect. If you have any questions, ask your teacher.

2) Be neat. Keep your work area clean. If you have long hair, pull it back so it doesn't get in the way. Roll up long sleeves. If you should spill or break something, or get cut, tell your teacher right away.

3) Watch your eyes. Wear safety goggles when told to do so.

4) Yuck! Never eat or drink anything during a science activity unless you are told to do so by your teacher.

5) Don't get shocked. Be sure that electric cords are in a safe place where you can't trip over them. Don't ever pull a plug out of an outlet by pulling on the cord.

6) Keep it clean. Always clean up when you have finished. Put everything away and wash your hands.

In some activities you will see these symbols. They are signs for what you need to do to be safe.

CAUTION

Be especially careful.

Wear safety goggles.

Be careful with sharp objects.

Don't get burned.

Protect your clothes.

Protect your hands with mitts.

Be careful with electricity.

UNIT A

LIFE SCIENCE

Plants and Animals

Chapter 1	**How Plants Grow**	**A2**
Chapter 2	**Types of Animals**	**A30**

Unit Project

Naturalist's Handbook

Observe the plants and animals that live around you. Make notes and sketches about your observations. Use tools like a hand lens, ruler, and thermometer to help you make your observations. Use reference materials to identify and describe the living things. Organize the information you find in a handbook to share.

Chapter 1

How Plants Grow

LESSON 1
What Do Plants Need? A4

LESSON 2
What Do Seeds Do? A10

LESSON 3
How Do Plants Make Food? A18

Science and Technology A24
People in Science A26
Activities for Home or School A27
CHAPTER REVIEW and TEST PREPARATION A28

Plants grow almost everywhere on Earth. Plants come in many sizes. Some are tall, like oak trees. Others are smaller, like rosebushes. But each plant has the same basic needs. These needs must be met in order for the plant to live and grow.

Vocabulary Preview

root
stem
leaf
seed
germinate
seedling
photosynthesis
chlorophyll

FAST FACT

Duckweed floats on the tops of ponds and quiet streams. It is the smallest flowering plant in the world.

Sizes of Plants	
Plant	Height
Duckweed	0.6 mm (0.02 in.)
Saguaro	15 m (50 ft)
Bamboo	30 m (100 ft)
Redwood	113 m (370 ft)

Frog in duckweed

FAST FACT

The largest cactus is the saguaro. It can be 50 feet tall and can weigh 14,000 pounds! In fact, it would take about 200 8-year-olds to match the weight of one giant saguaro.

LESSON 1

What Do Plants Need?

In this lesson, you can . . .

INVESTIGATE plant needs.

LEARN ABOUT how plants meet their needs.

LINK to math, writing, art, and technology.

Needs of Plants

Activity Purpose Plants need certain things to live and grow. In this investigation you will **observe** changes in plants. Then you will **compare** your observations to find out some of the things plants need to grow.

Materials

- 6 young plants
- 6 paper cups
- potting soil
- marker
- brown paper bag
- water
- ruler

Activity Procedure

1. Put the plants in the paper cups and add soil. Make sure all six plants have the same amount of soil. Label two of the plants *No Water*. Put these plants in a sunny window. (Picture A)

◀ The radishes you eat are roots.

Plants	Day 1	Day 3	Day 5	Day 7	Day 9	Day 11
No Water						
No Light						
Water and Light						

2 Label two plants *No Light*. Place these plants on a table away from a window. Water the plants. Then cover them with a paper bag.

3 Label the last two plants *Water and Light*. Water these plants. Put them in a sunny window.

Picture A

4 Every other day for two weeks, **observe** the plants. Check to make sure the plants labeled *No Light* and *Water and Light* have moist soil. Add enough water to keep the soil moist.

5 Make a chart like the one shown. On the chart, **record** any changes you **observe** in the plants. Look for changes in the color and height of each plant. (Picture B)

Picture B

Draw Conclusions

1. Which plants looked the healthiest after two weeks? Why do you think so?
2. Which plants looked the least healthy after two weeks? What was different for these plants?
3. **Scientists at Work** Scientists often **compare** observations to reach their conclusions. Compare your observations of the plants to tell what things plants need to grow. Make a list.

Process Skill Tip

When you **compare**, you tell how things are the same and how they are different. You may compare objects, events, or observations.

LEARN ABOUT

What Plants Need

FIND OUT
- four needs of plants
- how roots, stems, and leaves help plants live
- some different shapes and sizes of leaves

VOCABULARY

root
stem
leaf

These bluebonnets and Indian paintbrush flowers get what they need to live from nature. ▼

Plant Needs

Plants are living things. They live in places all over the world. Plants grow in deserts, in rain forests, and in your back yard. But no matter where they grow, all plants need the same things to live. As you learned in the investigation, these things include water and light. Plants also need soil and air.

Most plants live and grow without human care. They get what they need from the sun, the air, the rain, and the soil. Different kinds of plants need different amounts of these things. For example, a cactus needs very little water. Other plants, such as water lilies, need a lot of water. This is why the kinds of plants in one place may be very different from those in another place.

✓ **What four things do all plants need?**

The sun provides plants with light. ▼

Rain gives plants the water they need. ▼

Plant Parts

A tall oak tree looks very different from a daisy. A rosebush looks different from a dandelion. Yet all these plants have the same parts.

The **roots** of a plant are underground, where you often don't see them. A **stem** connects the roots with the leaves of a plant and supports the plant above ground. You may have seen a thin green stem on a flower. The thick, woody trunk of a tree is also a stem.

Leaves are plant parts that grow out of the stem. Most plants have many leaves.

✓ **How does water in the soil get to a plant's leaves?**

Roots, stems, and leaves are parts of a plant. They help the plant get what it needs to live.

The stem carries water from the roots to other parts of the plant.

Leaves take in the air and light a plant needs.

Roots hold a plant in the ground. They take in water and minerals from the soil.

Soil contains the minerals plants need. ▼

Leaf Shapes

Leaves help plants get the light and air they need. You may know that most leaves are green. But did you know that leaves have many shapes and sizes?

Many leaves, such as those from maple and oak trees, are wide and flat. Others, such as those from the jade plant, are small and thick. Leaves can be many different shapes. A leaf may have smooth or rough edges. Some leaf shapes may remind you of other things you have seen in nature. The way a leaf looks can help you tell what plant the leaf came from.

✓ **When you observe a leaf, what can help you tell what plant the leaf is from?**

▲ **The black oak leaf is long and wide, with rough edges.**

◄ **The leaf of the beech tree is oval and has rough edges.**

The leaf of the sweet-gum tree is star-shaped. Its edges are rough. ▼

The jade plant has leaves that are small and thick. The edges are smooth. ►

The ginkgo leaf is fan-shaped with smooth edges. ▶

Summary

Plants need water, light, soil, and air to live and grow. Roots, stems, and leaves help a plant get what it needs to live. Roots take in water from the soil. Stems carry this water to other parts of the plant. Leaves help the plant use light and air. Leaves have different shapes and sizes.

Review

1. Name four things a plant needs to live.
2. How does a plant take in water from the soil?
3. What plant part supports the leaves and branches of a tree?
4. **Critical Thinking** How might you care for an indoor plant?
5. **Test Prep** The plant parts that take in water and minerals from the soil are the —
   - A flowers
   - B roots
   - C leaves
   - D stems

LINKS

MATH LINK

Graphing Leaf Sizes Gather five to ten kinds of leaves. Use a ruler to measure the length of each leaf from the bottom of its stem to its tip. Make a bar graph to show the lengths of the leaves.

WRITING LINK

Informative Writing—Description Most states have a state tree and a state flower. What are these plants in your state? Write a description for your teacher of each.

ART LINK

Leaf Album Make drawings or crayon rubbings of your favorite kinds of leaves. Show the different shapes, sizes, and colors of the leaves. Label each drawing, and put it into a booklet.

TECHNOLOGY LINK

Learn more about plants and what they need by visiting the National Museum of Natural History Internet site.
www.si.edu/harcourt/science

 Smithsonian Institution®

LESSON

What Do Seeds Do?

In this lesson, you can . . .

 INVESTIGATE what seeds need to sprout.

 LEARN ABOUT why seeds are important.

LINK to math, writing, technology, and other areas.

 INVESTIGATE

Sprouting Seeds

Activity Purpose Cucumbers, carrots, and apples come from very different kinds of plants. But as different as these plants are, they all started as seeds. Seeds need certain things to sprout and grow. In this investigation, you will **observe** seeds to find out what they need to grow into new plants.

Materials

- 3 kinds of seeds
- paper towels
- 3 small zip-top bags
- scissors
- water
- tape
- hand lens

Activity Procedure

1. Start with a small amount of mixed seeds. Use size and shape to sort the seeds into three groups.

◀ These corn seedlings have roots, stems, and leaves.

Seeds	Days				
	1	2	3	4	5
Group 1					
Group 2					
Group 3					

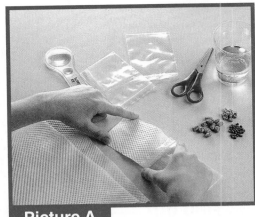
Picture A

2. Cut two paper towels in half. Fold the towels to fit into the plastic bags. Add water to make the towels damp. Do not use too much water or you will drown the seeds. (Picture A)

3. Put one group of seeds into each bag, and seal the bags. Label the bags *1*, *2*, and *3*. Tape the bags to the inside of a window.

4. Use a hand lens to **observe** the seeds every school day for 10 days. Use a chart like the one shown to **record** your observations. (Picture B)

Picture B

Draw Conclusions

1. What changes did you **observe** in the seeds?
2. How quickly did the changes take place in the different kinds of seeds?
3. **Scientists at Work** Scientists **observe** their investigations closely to get new information. How did observing the seeds help you understand more about seeds?

Investigate Further Repeat the investigation steps, but place the bags in a dark closet instead of in a window. **Predict** what will happen. **Record** your observations.

Process Skill Tip

When you **observe**, you use your senses to gather information. Observing sprouting seeds over time will help you understand more about how plants grow.

LEARN ABOUT

Growing Plants from Seeds

FIND OUT

- how plants reproduce
- what seeds need to sprout and grow
- four ways seeds are spread from one place to another

VOCABULARY

seed
germinate
seedling

What Seeds Are

Have you ever found seeds inside an orange or apple you were eating? Many plants reproduce, or make more plants like themselves, by forming seeds. The **seed** is the first stage in the growth of many plants.

Seeds may be large or small. They may be round, oval, flat, or pointed. Seeds may be many colors, too—they may even be striped. A seed looks very different from the plant that grows from it. But all seeds become plants that look like the plants they came from.

✓ **What does a seed do?**

After a sunflower seed germinates, a new sunflower plant will grow.

This cone from a pine tree produces many seeds.

A12

What Seeds Need

In the investigation, you observed that seeds began to grow into new plants when you gave them water. When the small plant breaks out of the seed, we say the seed **germinates**.

As a seed begins to sprout, a root grows from it. Next, the seed breaks open and a young plant, or **seedling**, appears. The seedling begins to form the parts it will need as an adult plant. The roots grow longer, and the stem begins to grow. Soon the seedling breaks through the soil. As the plant grows larger, leaves begin to form.

✓ **What do both seeds and plants need to grow?**

This summer squash is ready to eat. It is filled with seeds that could grow into new plants. ▼

How does the squash change as it grows from a seed to a young plant?

Some Plants Make Seeds

Plants that reproduce by seeds usually form more than one seed at a time. These seeds carry the information needed to grow into plants that look very much like the adult plant they came from.

Two groups of plants form seeds. In one group are the plants that have flowers. You may have seen flowers growing on bushes, on trees, or on small plants.

In the other group of seed-forming plants are plants that have cones. Some evergreen trees, such as pine trees, have cones. The cones have hard scales that protect the seeds under them.

Some plants can be grown from other plant parts. For example, if you place a leaf or a piece of a stem in water, it may form roots. Once roots form, you can put the plant part in soil, and it will grow into a new plant. The leaf or piece of stem used to grow the new plant is called a cutting. Underground plant parts such as tubers and bulbs can also be used to grow new plants.

✓ **Name five plants that form seeds.**

A bulb is a part of the stem. Many flowers, such as this tulip, grow from bulbs.

◀ Dandelions form many seeds that are easily blown by the wind.

The seeds in a watermelon are protected by its fruit. ▶

Seed Parts

Seeds may be many different shapes, sizes, and colors. But all seeds have the same parts. Seeds are like packages with tiny plants inside. The "package" is wrapped by the seed coat. Inside the seed coat is a seedling and food for the seedling.

✓ **What are the parts of a seed?**

The Parts of a Seed

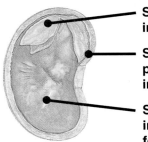

- **Seedling:** A seedling lives inside every seed.
- **Seed coat:** The seed coat protects the young plant inside the seed.
- **Stored food:** Most of the inside of a seed is stored food. The young plant uses the food to grow when the seed sprouts.

THE INSIDE STORY

Sizes of Seeds

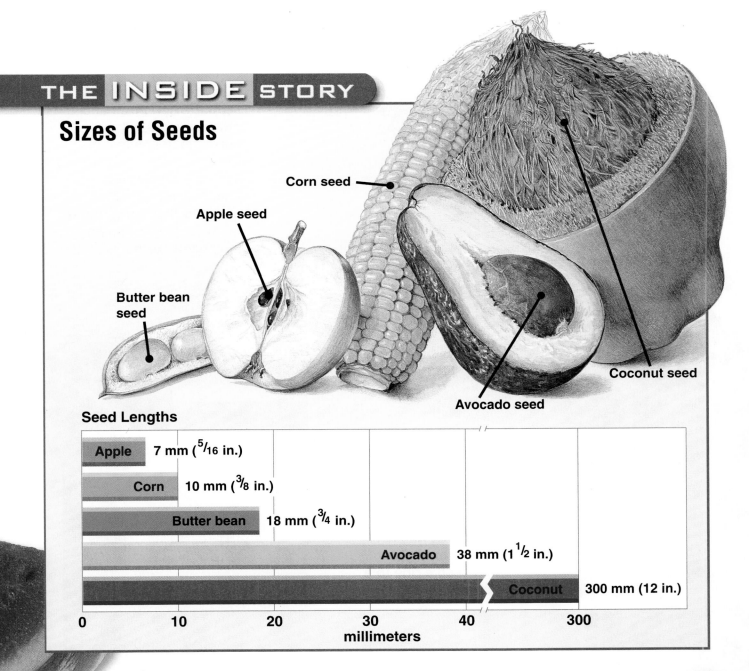

How Seeds Are Spread

Many seeds do not simply fall off the adult plants and sprout in the soil. Instead, they are scattered and moved to new places. Seeds are spread to new places in many ways. Some seeds are shot out of the adult plant like cannonballs from a cannon! Other seeds are spread by air, water, and animals.

✓ **What are some ways seeds are spread from one place to another?**

The fluffy parts of these milkweed seeds help them float away on the wind. ▶

Many berries contain seeds. When birds and other animals eat the berries, the seeds pass through their digestive systems unharmed. These seeds are later dropped onto the ground far away from the plant they came from.

◀ A cocklebur has hooks that grab onto fur or clothing. An animal or a person may give this seed a ride to a faraway place.

Water carries these mangrove seeds to new places. ▶

▲ Witch hazel plants shoot their seeds to spread them. This is called bolting.

Summary

Some plants form seeds to make new plants. Some plants also can be grown from plant parts. Seeds grow to look like the adult plants they came from. Although they may look different from one another, all seeds have the same parts and need water to sprout. Seeds are often spread to new places by air, water, and animals.

Review

1. What is a seedling?
2. What are some ways plants make new plants?
3. Describe some ways an animal can carry a seed from one place to another.
4. **Critical Thinking** If two seeds look alike, will they always grow into plants that look the same?
5. **Test Prep** What happens when a seed germinates?
   - A The seed breaks open and a seedling appears.
   - B The seedling breaks through the soil.
   - C Leaves begin to form.
   - D A root grows from the seed.

LINKS

MATH LINK

Counting Seeds Working with a partner, cut open an apple. Count the seeds inside. How many seeds in the whole class? Group the seeds by tens.

WRITING LINK

Narrative Writing—Story Write a creative story for a younger child about the life of a seed.

SOCIAL STUDIES LINK

Seeds Are Food Research the kinds of seeds people eat. Share with the class what you learn.

ART LINK

Fruity Posters Collect seeds from your favorite fruits. Use them to make a poster that tells about each fruit and its seeds.

TECHNOLOGY LINK

Visit the Harcourt Learning Site for related links, activities, and resources.
www.harcourtschool.com

LESSON 3

How Do Plants Make Food?

In this lesson, you can . . .

 INVESTIGATE what plants need to make their own food.

 LEARN ABOUT photosynthesis.

 LINK to math, writing, language arts, and technology.

Food Factories

Activity Purpose How do plants get food? They make it! To make food, plants must use the things around them. In this investigation, you will **observe** a plant making food. Then you will **infer** what plants need to make their food.

Materials

- scissors
- elodea
- dowel or pencil
- twist tie
- empty 0.5-L plastic bottle
- water
- brown paper bag
- watch or clock

Activity Procedure

1. **CAUTION** **Be careful when using scissors.** Use scissors to cut a piece of the elodea (el•oh•DEE•uh) as long as the bottle.

2. Wrap the elodea around the dowel. Use a twist tie to attach it to the dowel. (Picture A)

◀ Lemon trees make their own food.

A18

3. Put the elodea into the bottle, and fill the bottle with water. (Picture B)

4. Put the bottle in a place away from any windows. Cover the bottle with the brown paper bag. After 10 minutes, remove the bag. **Record** any changes you **observe.**

5. This time, place the bottle in bright sunlight and don't cover it with the brown paper bag. After 10 minutes, **observe** the bottle. **Record** any changes you observe.

Picture A

Draw Conclusions

1. Did the elodea and water in the bottle look different after Steps 4 and 5? If they did, tell how they were different.

2. What did you change between Steps 4 and 5? What remained the same in Steps 4 and 5?

3. **Scientists at Work** From what you **observed,** what can you **infer** about the bubbles you saw?

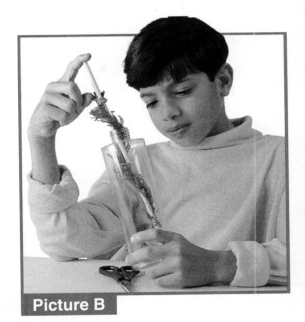

Picture B

Investigate Further Scientists often **measure** what happens in experiments. One way to measure what is happening in this experiment is to count the number of bubbles that appear. Put the bottle with the elodea in bright sunlight, and count the number of bubbles that appear in one minute. Then move the bottle out of the direct sun. Again count the number of bubbles that appear in one minute. How are the two measurements different? **Infer** why they are different.

> **Process Skill Tip**
>
> When you **infer,** you use what you have **observed** to form an opinion. That opinion is called an inference.

How Plants Make Food

FIND OUT

- how plants make their own food
- why plants need chlorophyll to make their food

VOCABULARY

photosynthesis
chlorophyll

Making Food

Like other animals, you cannot make food in your body. You must get your food by eating. Plants, however, can make their own food. This food-making process is called **photosynthesis** (foht•oh•SIN•thuh•sis). A process is a way of doing something.

The leaves of most plants are green. Plants get their green color from **chlorophyll** (KLAWR•uh•fil). Chlorophyll helps the plant use energy from the sun to make food. Plants need light to make food. They also need water and carbon dioxide. Carbon dioxide is a gas in the air.

The sun provides light energy to plants.

◀ A plant that gets enough light (and soil and water) grows into a healthy plant.

▲ A plant that does not get enough light does not grow well, even though it has soil and water. It may die.

THE INSIDE STORY

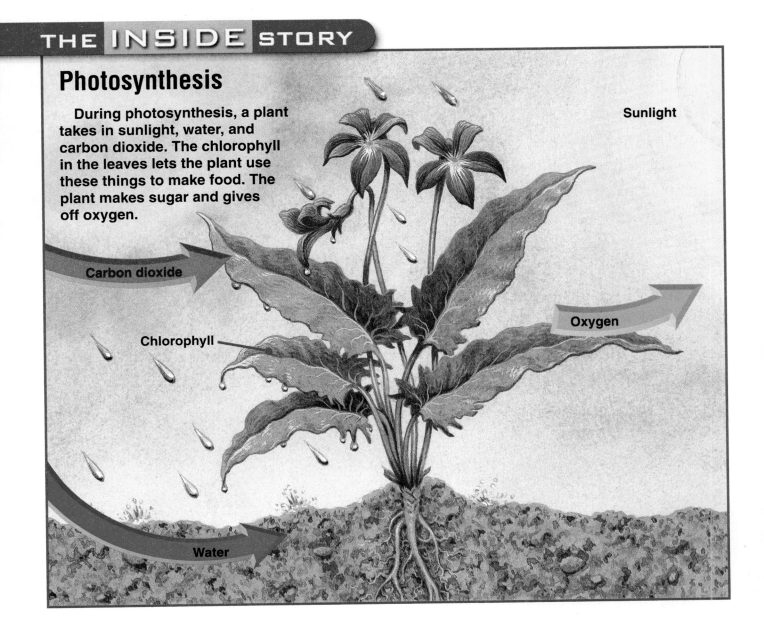

Photosynthesis

During photosynthesis, a plant takes in sunlight, water, and carbon dioxide. The chlorophyll in the leaves lets the plant use these things to make food. The plant makes sugar and gives off oxygen.

During photosynthesis, water and carbon dioxide combine inside a plant to make food. Sunlight provides the energy needed for this to happen. The food the plant makes is a kind of sugar. The plant uses some of the sugar right away and stores the rest.

During photosynthesis, plants also make oxygen. The oxygen is given off into the air through the plant's leaves. The bubbles you observed in the investigation were oxygen given off by the leaves into the water.

✓ **Name four things plants need for photosynthesis.**

How Plants Use Food

Plants use some of the food they make to grow larger. They also use it to make seeds. Some plants store food in their stems and roots so they can use it later. Other plants store sugar in fruits. This makes the fruit tasty to animals, who eat it and spread the seeds.

The stems, leaves, roots, seeds, fruits, and even flowers of plants are used as food by people and other animals.

✓ **What are three ways plants use the food they make?**

▲ Bananas grow in bunches called hands.

◄ People eat the red fruit of the strawberry plant.

◄ The stalks of celery plants contain fiber and water.

A potato is an underground stem that stores food made by the potato plant. ▼

▲ When you break open a pea pod you can see the pea seeds inside.

Summary

Photosynthesis is the process plants use to make their own food. Plants need chlorophyll, light, carbon dioxide, and water for photosynthesis. The sun provides the light energy that plants need to make food. Plants use the food they make to grow bigger and to make seeds. They store some of their food in roots, stems, and fruits. People and other animals eat many different plant parts.

Review

1. Describe what happens during photosynthesis.
2. Why are plants green?
3. How does chlorophyll help in photosynthesis?
4. **Critical Thinking** How does photosynthesis help make the food people and other animals eat?
5. **Test Prep** What do plants give off during photosynthesis?
   A carbon dioxide
   B water
   C oxygen
   D light

LINKS

MATH LINK

Plant Parts and Food Make a list of 20 foods you eat that come from plants. Identify how many of these foods come from roots, stems, leaves, fruits, seeds, and flowers. Make a bar graph from your list.

WRITING LINK

Informative Writing—Explanation Plants are like food factories. Use what you know about photosynthesis to write a paragraph for your teacher that explains how a plant is like a factory.

LANGUAGE ARTS LINK

Putting It Together The word *photosynthesis* is made up of the words *photo* and *synthesis*. Use a dictionary to find out what these words mean.

TECHNOLOGY LINK

To learn more about how people use the food that plants store, watch *Grocery Garden* on **Harcourt Science Newsroom Video.**

SCIENCE AND TECHNOLOGY

Drought-Resistant Plants

Different kinds of plants live in different kinds of places. Water plants need lots of water. Plants that live in the desert need very little water. Plants that need little water are called drought-resistant plants.

What Makes Plants Drought Resistant?

Drought-resistant plants have traits that allow them to live successfully with little water. One trait is gray or white leaves or bark. Their light color reflects the heat of the sun rather than absorbing it. This keeps the plant cooler, so it doesn't dry out as much. Another trait these plants have is narrow leaves, which help the plant keep moisture in.

Many drought-resistant plants bloom early in the spring, when there is plenty of water. They develop seeds before the hot, dry months.

Some plants have underground bulbs or enlarged root systems that store food. When dry weather arrives, the plant has plenty of food it can use.

Who Needs Drought-Resistant Plants?

Some parts of the United States get little rain. Many of these places are warm and get lots of sun. More and more people are deciding to live in these warm, sunny places. These people often try to grow gardens with the kinds of plants they had where they came from. These plants need more water than they get naturally in dry areas, so people have to water them a lot. But this uses water that is needed for other things, such as farms, animals, and people.

One way to have pretty plants without using much water is to use a process called Xeriscaping (ZIR•uh•skayp•ing). Xeriscaping is the use of water-saving methods and drought-resistant plants in gardens and yards.

How Does Xeriscaping Work?

First the gardener analyzes the soil and the amount of water in the garden area. Then the gardener chooses plants that can grow well in those conditions. Often the gardener chooses plants that are native to the area because they have adapted to the local climate. The gardener puts a layer of mulch, such as leaves or wood chips, around each plant to help hold water in the ground. Xeriscaping saves water costs for the gardener and conserves water for the whole community.

Think About It

1. How can planting drought-resistant plants help the environment?
2. What are some native plants in your area?

WEB LINK:
For Science and Technology updates, visit the Harcourt Internet site.
www.harcourtschool.com

Careers: Agricultural Extension Agent

What They Do Extension agents often work for state or local governments. These people offer help and information to farmers, gardeners, and teachers. They often travel to different areas to teach and train people in new technology used in agriculture.

Education and Training An extension agent must study agriculture and animal science. Training in computers and agricultural technology will be required for the agents of the future.

PEOPLE IN SCIENCE

George Washington Carver
AGRICULTURAL CHEMIST

"I literally lived in the woods. I wanted to know every strange stone, flower, insect, bird, or beast."

How many ways can you think of to use a peanut? George Washington Carver found more than 300!

After completing college, Carver was invited to teach at the Tuskegee Institute in Alabama. For almost 50 years, he taught there and conducted experiments with plants. He experimented to solve many problems of farming. The problems included poor soil, amounts of sunlight and moisture, plant diseases, and ways of reproducing plants.

Carver persuaded farmers to rotate their crops so the soil would not lose all its nutrients. One year the farmers could plant cotton or tobacco, but the next year they planted peanuts or sweet potatoes. The land produced so many crops that Carver had to find new uses for them. Besides his work with peanuts, he found more than 100 uses for sweet potatoes and 75 for pecans.

Think About It
1. George Washington Carver was interested in plants from his boyhood. Think about the things that interest you now. Which ones do you think will interest you when you grow up?
2. What qualities do you think a plant scientist ought to have?

Some of the plants that Carver worked with ▼

ACTIVITIES FOR HOME OR SCHOOL

What Stems Do

How does water move through a stem?

Materials
- 3 1-L plastic bottles
- water
- food coloring
- 3 freshly cut white carnations
- scissors

Procedure

1. Fill the plastic bottles with water. Add a few drops of food coloring. Put a different color in each bottle.

2. **CAUTION** Be careful when using scissors. Trim the end off the stem of each flower. Place one flower in each bottle.

3. Keep the flowers in the bottles overnight. Then observe the flowers.

Draw Conclusions

Explain your observations. Make a bouquet of flowers with the colors you like best.

Growing Plants

Can plant parts be used to grow a new plant?

Materials
- 1 plant with many stems and leaves
- scissors
- 1-L plastic bottle
- water
- ruler

Procedure

1. **CAUTION** Be careful when using scissors. Cut off a 15-cm piece of the plant.

2. Fill the plastic bottle with water. Place the cut part of the plant in the bottle.

3. Place the plant in bright sunlight.

Draw Conclusions

Observe the plant for ten days. Record any changes you observe.

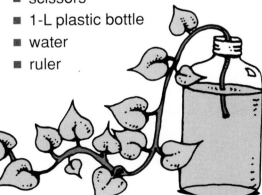

Chapter 1 Review and Test Preparation

Vocabulary Review

Use the terms below to complete the sentences 1 through 7. The page numbers in ( ) tell you where to look in the chapter if you need help.

roots (A7) **germinate** (A13)
stem (A7) **photosynthesis** (A20)
leaves (A7) **chlorophyll** (A20)
seeds (A12)

1. Many new plants grow from ____.
2. The plant parts that grow out of stems are ____.
3. ____ gives plants their green color.
4. A seed needs water to ____.
5. The underground parts of a plant that take in water from the soil are the ____.
6. The food-making process of plants is ____.
7. A ____ connects the roots and leaves of a plant.

Connect Concepts

Use the terms from the Word Bank to complete the concept map.

wind   **chlorophyll**   **water**   **oxygen**   **animals**

How Plants Grow		
What Plants Need	**What Seeds Do**	**How Plants Make Food**
Roots help a plant take in water from the soil.	Many plants can form seeds. Seeds need air and water to sprout. Seeds are spread in many ways.	Plants make their own food by a process called photosynthesis.
Leaves take in carbon dioxide from the air. They give off 8. ____.	Three ways seeds are spread are by 9. ____, 10. ____, and 11. ____.	12. ____ in plants helps take in light.

A28

Check Understanding

Write the letter of the best choice.

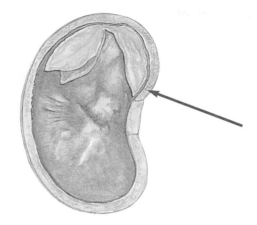

13. Look at the picture. Which part of the seed is the arrow pointing to?
    A stored food   C seed coat
    B young plant   D stem

14. Some plants are grown from bulbs. A bulb is a part of the —
    F root   H stem
    G seed   J leaf

Critical Thinking

15. What would happen to a farmer's crop if there were little or no rain for a long period of time?

16. How can a dandelion growing in the schoolyard be a parent to young dandelion plants growing miles from school?

17. If a plant reproduces by bulbs, does it make seeds, too? Explain.

Process Skills Review

18. You are growing two plants. You want to test one plant to find out how sunlight affects its growth. You will provide the other plant with everything it needs for growth. Explain how you will **observe** the plants and **compare** your results.

19. What can **observing** tree leaves help you learn?

Performance Assessment

Design a Garden

With a partner, design a garden to grow your favorite kinds of plants. Make a list of all the things you will need for your garden.

Chapter 2
Types of Animals

LESSON 1
What Is
an Animal? A32

LESSON 2
What Are
Mammals
and Birds? A40

LESSON 3
What Are
Amphibians,
Fish, and
Reptiles? A48

Science
Through Time A58

People in
Science A60

Activities for Home
or School A61

CHAPTER REVIEW
and TEST
PREPARATION A62

Vocabulary Preview

inherit
trait
mammal
bird
amphibian
gills
fish
scale
reptile

How many different kinds of animals can you name? You probably listed animals such as cats, dogs, cows, and fish. But there are many more. Even though there are many kinds of animals, they all need the same types of things to live and grow.

FAST FACT

It's a trick! These butterfly fish have spots on their tails that look like eyes. The "eyes" scare away larger fish that might eat them.

FAST FACT

Types of Animals on Earth

75% insects

25% all other animals

Swat! Slap! Mosquitoes bite. Bees sting. Of every 100 types of animals in the world, 75 are insects.

LESSON 1

What Is an Animal?

In this lesson, you can . . .

INVESTIGATE animal homes.

LEARN ABOUT how animals meet their needs.

LINK to math, writing, literature, and technology.

◀ Ladybugs get food from this bluebonnet.

Animal Homes

Activity Purpose Everyone needs a place to live—including animals. Animals live in many kinds of places that give them the things they need. In this investigation you will **observe** animal homes and use the homes to **classify** the animals.

Materials
- Animal Picture Cards

Activity Procedure

1. Select six Animal Picture Cards or use the pictures on page A32. As you **observe** the cards, pay close attention to the types of homes the animals live in. (Picture A)

2. Describe each animal home you **observe**. **Record** your descriptions.

3. With a partner, discuss the different types of animal homes shown. Talk about the ways the animal homes are alike and the ways they are different. Then **classify** the animals by the types of homes they live in.

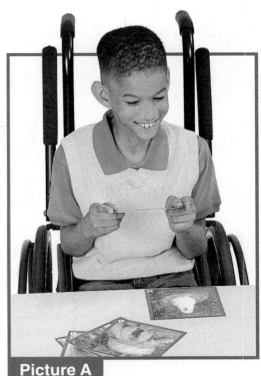

Picture A

Draw Conclusions

1. **Compare** two of the animal homes you observed. Tell how each home helps protect the animal that lives there.

2. What did you **observe** about the home of a Canada goose and the home of an albatross?

3. **Scientists at Work** Scientists **classify** animals into groups based on what the animals have in common. How many groups did you classify the animals into? What were the groups?

Investigate Further Study the Animal Picture Cards again. This time, look at the body covering of each animal. Describe each covering. How can you use body coverings to **classify** the animals?

Process Skill Tip

When you **classify** things, you put them into groups. You put things into a group because they have something in common. For example, you could put into one group the animals that build nests.

LEARN ABOUT

Animals and Their Needs

FIND OUT

- what animals need so they can live
- how animals' bodies help them meet their needs
- how animals get their traits

VOCABULARY

inherit
trait

What Animals Need

Have you ever cared for a pet? If so, you know that a pet needs food and water. A pet also needs shelter, or a place to live. Wild animals have the same needs as pets. The difference is that wild animals must meet their needs on their own.

As you learned in the investigation, different animals make their homes in different surroundings. From their surroundings, they get all the things they need. This is something all animals have in common.

✓ **Name three things animals need.**

This reef is made up of tiny living animals called corals. Other animals, such as the dolphin shown here, look for food in the reef and use the reef for shelter. ▼

Animals Need Air

Have you ever thought about the air around you? Air contains a gas called oxygen. Animals need oxygen to live.

Animals that live on land, such as giraffes, have lungs that get oxygen from the air. Insects get oxygen from the air through tiny holes in their bodies. Many water animals, such as fish, get their oxygen from water. Other water animals, such as whales, must come to the surface and breathe air to get oxygen.

✓ **Name two ways animals get oxygen.**

▲ An alligator comes to the water's surface to breathe the air it needs.

Animals Need Water

The bodies of all animals contain water. Every day some of this water leaves the animals' bodies. For example, when an animal pants or sweats, it loses some water from its body. The animal must replace this water to stay alive.

Most animals get the water they need by drinking. The water they drink may come from puddles, streams, rivers, ponds, or lakes. Other animals get most of their water in the foods they eat.

✓ **How do animals get the water they need?**

At this water hole on the African plains, animals drink the water they need. ▼

Animals Need Food

All animals need food. Food gives animals the materials they need so they can grow and stay healthy. Animals also get energy from food.

Unlike plants, animals cannot make their own food. Instead, they get their food by eating plants or other animals. Some animals eat only plants. Some eat only animals. Some eat both.

How do animals get the food they need? Many of them have body parts that help them get their food. For example, an elephant uses its trunk to grab leaves from trees. A hawk can use its sharp claws to catch a mouse.

✓ **What do animals eat?**

▲ This chameleon uses its tongue to catch insects.

Brown bears live in forests, where they eat many kinds of plants and animals. They often eat fish they catch in streams. They also eat grass and other plants.

A panda eats only bamboo plants. ▼

THE INSIDE STORY

Beavers Build Shelters

1. Beavers cut down trees to build a dam in a stream.

2. The dam makes a pond. In the pond, the beavers build their home, called a lodge.

3. To build their lodges, beavers use trees they have cut down and rocks that they cover with mud.

4. Young beavers, called kits, stay warm and dry inside the lodge.

Animals Need Shelter

Most animals need shelter, or a place to live. Shelters protect animals from other animals and from the weather. Some birds build shelters called nests high in tree branches. They build their nests out of twigs, grass, and mud. Other animals, such as deer mice, build their homes in hollow logs or in spaces under rocks. Turtles use their own hard shells as their homes. Many other animals dig tunnels and make their homes in the ground.

✓ **Why do animals need shelter?**

Animal Traits

Jellyfish, polar bears, and snakes don't look much alike, but they are all animals. Different animals have many different shapes and sizes. They also have many different body parts. For example, a bird has wings and feathers. A lion has paws and fur. All these features are important to the way an animal lives.

How do animals get their features? Young animals inherit their features from their parents. **Inherit** (in•HAIR•it) means "to receive from parents." The body features an animal inherits are called **traits**. Traits also include some things that animals do.

✓ **What are traits?**

▲ A sea horse is a type of fish. This male sea horse has the trait of carrying its young inside a pouch.

These young cheetah cubs will grow to look like their parents. They will stay with their mother for several years and will learn to hunt for their food. ▼

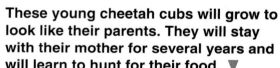

▲ A mother duck watches her young ducklings carefully.

▲ These young snakes will grow to look very much like the adult snake.

Summary

Animals need air, water, food, and shelter. Different animals have many different shapes and sizes and many different body parts. These traits help the animals get the things they need. All animals inherit their traits from their parents.

Review

1. How do animals that live in water get air?
2. Describe how beavers change their surroundings to meet their needs.
3. What are four things that animals need?
4. **Critical Thinking** Why will a young lion cub grow to look like an adult lion and not like a sea horse?
5. **Test Prep** Which of the following is **NOT** a need of all animals?

   A food   C air

   B water   D soil

LINKS

MATH LINK

How Many? Suppose that there are 15 duck families that live at a pond. If each of the families has 8 ducklings, how many ducks would live at the pond?

WRITING LINK

Informative Writing— Compare and Contrast Birds make many sounds. Listen to birds that live in your area. If possible, record their sounds with a sound recorder. Write a paragraph for your teacher that compares and contrasts the bird songs.

LITERATURE LINK

A Wolf Pup Diary You can learn about the life of a young wolf by reading *Look to the North: A Wolf Pup Diary* by Jean Craighead George.

TECHNOLOGY LINK

Learn more about animals by visiting the Smithsonian Internet Site.
www.si.edu/harcourt/science

Smithsonian Institution®

LESSON 2

What Are Mammals and Birds?

In this lesson, you can . . .

INVESTIGATE how fur helps animals.

LEARN ABOUT the traits of mammals and birds.

LINK to math, writing, art, and technology.

Fur Helps Animals

Activity Purpose When it's cold outside, you might put on a jacket or a sweater to keep warm. Animals can't do that. In this investigation you will **use a model** to find out how fur helps keep animals warm.

Materials
- glue
- 2 metal cans
- cotton batting
- hot water
- 2 thermometers
- classroom clock

Activity Procedure

1. Make a chart like the one shown.

2. Spread glue around the outside of one can. Then put a thick layer of cotton around the can. Wait for the glue to dry. Then use your fingers to fluff the cotton. (Picture A)

◀ The fur on this deer keeps him warm all winter long.

Time	Water Temperature in Can with Cotton	Water Temperature in Can Without Cotton
Start		
10 min		
20 min		
30 min		

3 **CAUTION** **Be careful with the hot water. It can burn you.** Your teacher will fill both cans with hot water.

4 Place a thermometer in each can, and **record** the temperature of the water. (Picture B)

5 Check the temperature of the water in each can every 10 minutes for a period of 30 minutes. **Record** the temperatures on the chart.

Picture A

Picture B

Draw Conclusions

1. In which can did the water stay hot longer? Why?
2. How is having fur like wearing a jacket?
3. **Scientists at Work** Scientists often **use a model** to study things they can't observe easily. In this investigation, you made a model of an animal with fur. Why was using a model easier than observing an animal?

Investigate Further How do you think your results would be different if you used ice-cold water instead of hot water? Make this change, and do the investigation again.

Process Skill Tip

It would be hard to measure the temperature of a real animal to find out how fur helps it stay warm. **Using a model** helps you learn about animal fur.

Mammals and Birds

FIND OUT

- four traits of mammals
- five traits of birds

VOCABULARY

mammal
bird

Mammals

In the investigation you learned that fur can help an animal stay warm. Animals that have fur or hair are called **mammals** (MAM•uhlz). Horses, cows, and dogs are all mammals.

Mammals use lungs to breathe. Mammals that live in water, such as whales, also breathe with lungs. But these mammals must come to the surface of the water to breathe the air they need.

Most mammals give birth to live young. A cat gives birth to many kittens at one time. Before the kittens

THE INSIDE STORY

Keeping Warm

A polar bear's fur looks white, but it is really clear. The fur looks white because it reflects sunlight.

The skin of a polar bear is black. The black skin takes in the heat from the sun. Polar bears have a thick layer of fat under their skin. This fat helps keep a polar bear warm.

are born, the mother cat carries them inside her. After the kittens are born, they feed on milk made by their mother's body. Feeding their young with milk from the mother's body is another trait of all mammals.

When a mammal is born, it cannot care for itself. It must be sheltered and fed by its mother. The milk gives them what they need to grow and stay healthy.

Most mammals learn from their parents how to care for themselves. The parents often teach their young how to find food. In time a young mammal learns what it needs to know to live on its own.

✓ **What are four traits mammals inherit from their parents?**

▲ These puppies drink milk from their mother to get the food they need.

◀ Gorillas often live in large families. All the adults help care for the young.

Types of Mammals

There are many types of mammals. Most of them have the four traits you read about. Some also have other traits, such as trunks, pouches, or wings. Mammals are often placed into groups based on traits they share. Some of these groups and their traits are shown on this page.

✔ **Name three traits used to place mammals in groups.**

▲ Bats are the only mammals that fly.

A koala is one of a few mammals that carry their young in a pouch. ▼

▲ This spiny echidna (ee•KID•nuh), an anteater, has fur and lungs. It is a mammal, but it does not give birth to live young. It lays eggs.

This orangutan (oh•RANG•oo•tan) is a primate. Primates are mammals that can use their hands to grasp objects. ▼

Whales are mammals that live in water. They have very little hair. This helps them glide easily through the water. ▶

Birds

Birds are animals that have feathers, two legs, and wings. Most birds use their wings for flying. Some birds, such as penguins, cannot fly. But like other birds, they still have feathers and wings.

Like mammals, birds have lungs for breathing air. Many birds also care for their young for a while after the young are born. Unlike most young mammals, young birds hatch from eggs.

Feathers cover most of a bird's body. But not all feathers are the same. Some feathers help keep a bird warm. Other feathers help birds fly. For example, the wing feathers of many birds have a shape that helps them fly.

✓ **What are five traits of birds?**

▲ A weaverbird uses leaves to make a hanging nest. The mother bird lays her eggs on soft grass placed inside the nest.

Types of Birds

There are many types of birds. Like mammals, birds are grouped together because of traits they share. The most common traits used for grouping birds are beak shape and foot shape.

Beak shape can be used to tell what kind of food a bird eats. For example, wading birds have long beaks that help them catch fish and dig small animals from the mud.

Foot shape can be used to tell where a bird lives. For example, wading birds have long toes that keep them from sinking into the mud. For some birds foot shape is also important in getting food.

✓ **What can you tell about a bird from the shape of its beak or feet?**

▲ A barn owl catches other animals for food. The feet of a barn owl are useful for catching animals such as mice.

A cardinal uses its beak to eat seeds. It uses its feet to hold on to branches. ▶

The purple gallinule (GAL·ih·nool) lives in marshes and swamps. It has long, thin toes that help it walk on lily pads and other plants in the water. ▼

The great blue heron is a wading bird. Its beak has a shape that is useful for catching fish. ▶

Summary

Mammals are animals that have fur or hair and breathe with lungs. Most mammals also give birth to live young and feed their young with milk from the mother's body. Birds are animals that have feathers, two legs, and wings. Like mammals, birds breathe with lungs. Unlike most mammals, birds lay eggs from which their young are hatched.

Review

1. Name a mammal that lives in the water. How does it breathe?
2. How does the spiny echidna differ from most other mammals?
3. What two features of birds are most often used to classify them? Why?
4. **Critical Thinking** A bat can fly, but a bat is a mammal. What traits do you think bats have that make them mammals instead of birds?
5. **Test Prep** Which trait is shared by birds and mammals?
   A have feathers
   B have fur
   C breathe with lungs
   D give birth to live young

LINKS

MATH LINK

Interpret Graphs Look at the graph on page A45. Which birds have the longest tails? If the length of the tails were added to the length of the bodies in the graph, which bird would be the longest?

WRITING LINK

Informative Writing— Classification Go bird-watching in your schoolyard, a park, or in your back yard. If possible, use binoculars. Write a description of each bird to share with your classmates. Identify as many birds as you can.

ART LINK

Animal Tracks You can identify many animals by the tracks they make with their feet. Make a collage of tracks. Label each track with the animal's name.

TECHNOLOGY LINK

Learn more about mammals and birds by investigating *Whose Tracks Are These?* on **Harcourt Science Explorations CD-ROM.**

LESSON 3

What Are Amphibians, Fish, and Reptiles?

In this lesson, you can . . .

INVESTIGATE how frogs change as they grow.

LEARN ABOUT amphibians, fish, and reptiles.

LINK to math, writing, technology, and other areas.

INVESTIGATE

From Egg to Frog

Activity Purpose Frogs lay eggs in the water. When a young frog hatches, it can live only in water. But as the young frog grows, its body changes. The changes get it ready to live on land. In this investigation you will **observe** these changes in real frogs.

Materials

- gravel
- aquarium
- water
- ruler
- water plants
- rock
- tadpoles
- dried fish food

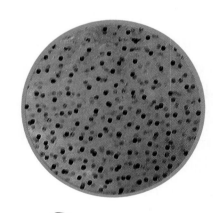

◄ These frog eggs were laid underwater. They will hatch into tadpoles.

◄ Unlike frogs you may know about, this red-eyed tree frog makes its home in trees.

Activity Procedure

Picture A

1. Put a layer of gravel on the bottom of the aquarium. Add 12 cm to 15 cm of water.

2. Float some water plants on top of the water, and stick others into the gravel. Add the rock. It should be big enough so that frogs can sit on it later and be out of the water. (Picture A)

3. Put two or three tadpoles, or young frogs, in the water. Put the aquarium where there is some light but no direct sunlight.

4. Feed the tadpoles a small amount of dried fish food once a day. Add fresh water to the aquarium once a week.

5. **Observe** the tadpoles every day. Once a week, make a drawing of what they look like.

Draw Conclusions

1. What changes did you see as the tadpoles grew?
2. When the tadpoles began to climb out of the water, what did their bodies look like?
3. **Scientists at Work** Scientists **record** what they **observe**. How did recording your observations help you learn about the growing tadpoles?

Investigate Further How important were the water plants to the growth of the tadpoles? Plan an investigation to answer this question.

> **Process Skill Tip**
>
> When you **observe**, you use your senses of sight, hearing, smell, and touch. Then you **record**, or write down, your observations.

LEARN ABOUT

Amphibians, Fish, and Reptiles

FIND OUT

- four traits of amphibians
- four traits of fish
- three traits of reptiles

VOCABULARY

amphibian
gills
fish
scales
reptile

Amphibians

In the investigation, you observed that a tadpole lives in water. But as the tadpole grows into a frog, it spends more time out of the water. Frogs are amphibians. **Amphibians** (am•FIB•ee•uhnz) are animals that begin life in the water and move onto land as adults.

Amphibians lay eggs in the water. The eggs stay there until they hatch. Young amphibians live in the water, just as tadpoles do. Most adult amphibians live on land.

Amphibians have moist skin. But most amphibians still stay close to water.

✓ **Name three traits of an amphibian.**

◀ A salamander moves its body from side to side to help its legs move forward.

Newts live in water for part of the year. Then they live on land. ▼

▲ Toads live most of their adult lives on land. Unlike some other amphibians, toads have rough, bumpy skin.

THE INSIDE STORY

Frog Metamorphosis

A frog changes as it grows from an egg to an adult. The changes it goes through are called *metamorphosis* (met·uh·MAWR·fuh·sis).

1. A frog lays many eggs at one time. The eggs are covered by a jellylike coating.

2. Newborn tadpoles have gills for breathing in water. They also have a tail but no legs.

3. As a tadpole grows, lungs begin to form. Back and front legs begin to grow. These parts allow an adult frog to live on land.

4. Once the lungs work, the gills and the tail disappear. The adult frog is now ready to live on land.

Frogs Grow and Change

As you saw in the investigation, young frogs look very different from adult frogs. Young frogs hatch from eggs and begin life in the water. They breathe with gills. **Gills** are body parts that take in oxygen from the water.

As they grow, young frogs change. In time, they form lungs. Once they have lungs, their gills begin to disappear. Then they develop other body parts that help them live on land. Adult frogs spend most of the year on land near water.

✓ **What body parts do young frogs have that adult frogs do not have?**

Fish

Fish are animals that live their whole lives in water. Like young amphibians, fish have gills. The gills are on the sides of a fish's head. The gills take in oxygen as water moves over them.

Most fish are covered with scales. **Scales** are small, thin, flat plates that help protect the fish.

Different fish have many different shapes and sizes. Like other animals, some fish eat plants and others eat animals. Most fish lay eggs, but some fish give birth to live young.

✓ **What are two traits of fish?**

▲ This Guadeloupe bass has a skeleton made of bone.

▲ This ray has a cartilage skeleton just like a shark, but its body is flat.

Sharks do not have bones. Instead, their skeletons are made of a softer material called cartilage (KAR•tuh•lij). ▼

The Bodies of Fish

A fish's body is just right for its life in water. Most fish have body shapes that allow them to move easily in water. The smooth scales that cover fish also help them glide through water.

Fish have fins. They use their fins to move forward and backward. The tail fin moves from side to side and helps a fish move forward. Other fins help the fish turn in different directions.

✓ **What are three ways a fish's body is just right for life in water?**

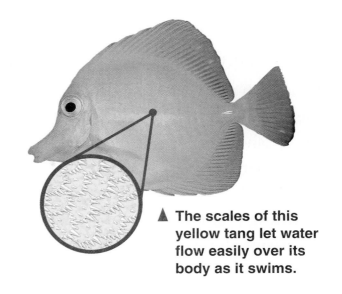

▲ The scales of this yellow tang let water flow easily over its body as it swims.

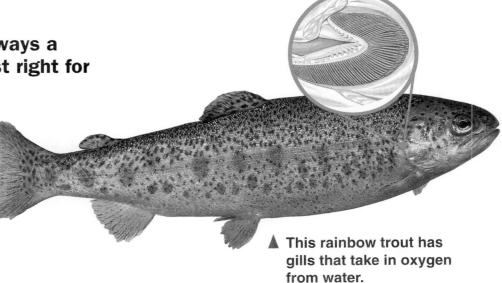

▲ This rainbow trout has gills that take in oxygen from water.

▲ The fins on this goldfish help it swim and turn its body through the water.

Fish Young

Young fish are hatched from eggs. Some fish carry their eggs inside their bodies until the eggs hatch. But most fish lay their eggs in water. They lay many eggs at a time.

Some fish care for their eggs by guarding them. But most fish leave their eggs alone after they are laid. Other animals eat many of the eggs. The young fish that do hatch are very small. But they must find food and avoid enemies without help from their parents.

✓ **What are two ways fish produce young?**

▲ This stickleback fish builds a nest to lay its eggs in.

This cardinal fish carries its eggs in its mouth until they are ready to be hatched. ▼

Reptiles

Reptiles (REP•tylz) are land animals that have dry skin covered by scales. Because they live mostly on land, reptiles use lungs to breathe air. Reptiles that spend a lot of time in water must come to the surface to breathe air. A crocodile, for instance, often stays in water. But it stays near the surface so its nose and eyes are above the water. This allows the crocodile to breathe and see.

Many reptiles hatch from eggs laid on land. The eggs have a tough, leathery shell. Other reptiles are born live. Either way, most of the young are able to meet their needs as soon as they are born.

Reptiles are found almost everywhere on Earth except for the coldest places. Many live in warm, wet tropical rain forests or in hot, dry deserts.

✓ **What are three traits of reptiles?**

▲ The Eastern box turtle hatches from eggs laid on land.

The turtle has a hard shell that protects most of its body. The turtle has scales on its legs and tail. ▼

Types of Reptiles

There are three main groups of reptiles. Lizards and snakes are in one group. Their bodies have rows of scales that overlap. Most lizards live in very warm places. They have four legs and long tails. Snakes don't have legs. They move by pushing their bodies against the ground.

Alligators and crocodiles are another reptile group. They live in water a lot of the time. They come out of the water to sun themselves.

Tortoises and turtles make up the third group of reptiles. They are the only reptiles that have shells. Tortoises live on land. Turtles live in water.

✓ **What are the three groups of reptiles?**

▲ The boa is a large snake. Like other snakes, it sheds its scaly skin and grows new skin.

▲ The Australian frilled lizard uses a fan-shaped body part to scare away enemies.

This is an American crocodile. A crocodile can lie very still in the water, with only its eyes and nose above the water. This helps the crocodile catch animals that can't see it. ▼

▲ This is a Galápagos tortoise. Like turtles, tortoises have hard shells that protect them from enemies.

Summary

Amphibians are animals that begin life in water, change in form, and then live on land. Fish live in the water, use gills for breathing, and have body parts that help them swim. Reptiles are animals that are covered with scales.

Review

1. What happens during the metamorphosis of a frog?
2. List three features that help fish live and move in water.
3. What do gills do?
4. **Critical Thinking** Why do many amphibians stay near the water for their whole lives?
5. **Test Prep** Which trait is shared by most fish, amphibians, and reptiles?
   A fins   C eggs
   B scales   D legs

MATH LINK

Using Graphs Take a survey of the kinds of amphibians, fish, and reptiles that people keep as pets. Use a computer graphing program such as *Graph Links* to make a bar graph to show your results.

WRITING LINK

Narrative Writing—Story What if you were a frog? Write a story for a younger child telling how you change from an egg into an adult.

LITERATURE LINK

Verdi To learn more about a snake called a python, read *Verdi* by Janell Cannon.

ART LINK

Collage Make a collage that shows adult animals with their young.

TECHNOLOGY LINK

Learn about endangered reptiles by watching *Endangered Animals* on the **Harcourt Science Newsroom Video.**

SCIENCE THROUGH TIME

Discovering Animals

No one knows how many kinds of animals there are. New kinds are found every year. There are more than one and one-half million kinds of animals! More than one million kinds of animals are insects. Some scientists think there could be as many as 50 million kinds of animals!

Grouping Animals

People have been observing animals for centuries. New animals are discovered as scientists and explorers go to new places. One of the first people to observe animals and to write about them was Aristotle. He lived about 350 B.C. He divided animals into two groups. One was made up of animals with backbones and red blood. This group

The History of Animal Discovery

350 B.C.
Aristotle develops a classification system for animals.

1400–1600
Age of exploration—many animals are discovered.

1590
The microscope is invented.

1600s
John Ray discovers that whales are mammals.

included horses, cats, dogs, and oxen. The other group did not have backbones or red blood.

Since then, scientists have discovered that most animals do not have backbones. Clams, spiders, ants, sponges, worms, jellyfish, and squids are all part of this group.

Your senses can mislead you when you study animals. A whale looks like a fish, and it lives in the water. Not until the 1600s did an English scientist, John Ray, observe whales closely. He learned that they are mammals, just as humans are.

Discovering New Animals

The microscope was invented during the late 1500s. Using this tool, scientists began to discover living things made up of only one cell. In 1665 Robert Hooke published a book of drawings of biological specimens viewed through a microscope. Scientists today use microscopes to study all kinds of cells. These studies help scientists understand how animals are related to one another.

When explorers during the 1500s and 1600s traveled to new places, they returned home with animals that had never been seen before. They brought parrots and other brightly colored birds from tropical areas such as South America. They also brought many different kinds of monkeys from those areas. A giraffe from East Africa was sent to China. All of these animals were studied carefully by scientists.

People continue to find and study new animals. Modern explorers search the oceans for unknown species. We still know very little about animals living at the bottom of the ocean. Thermal vents are places where heat comes from the ocean floor. Many strange creatures live near these thermal vents.

Scientists will continue to explore unfamiliar places, such as rain forests, ocean floors, and volcanoes. Each new place will probably have animals we have never seen before.

Think About It

- How have microscopes helped scientists learn about animals?

1665
Robert Hooke publishes a collection of his drawings.

1977
Animals are discovered living near thermal vents deep in the ocean.

PEOPLE IN SCIENCE

Rodolfo Dirzo
TROPICAL ECOLOGIST

"I am interested in... the loss of animals in tropical ecosystems."

Growing up in Mexico, Rodolfo Dirzo used to watch bugs. That early interest led him to study snails and slugs far away in Wales. After completing his education, he returned to Mexico. He has taught for many years at the Organization for Tropical Studies. He does research on tropical forests. Teaching is one of Dirzo's main interests. He especially wants to interest Latin American students in ecology.

Besides teaching, Dirzo studies two different forests in Mexico. In one, the forest has not been disturbed. There are many kinds of plants and animals. In the other, some of the forest has been cut down. As a result, many of the animals that would be expected to live there are no longer there. Dirzo compares the two places. He uses his imagination and his training in science to describe what happens to a habitat that has changed. Without animals to help spread seeds and to trample down the vegetation, the forest plants may change. Dirzo expects that there will be fewer kinds of trees without the animals to help provide places for many different plants to grow.

Think About It

1. What might cause animals to leave a forest?
2. How does comparing the two forests help Dirzo analyze his data?

Rain forest

ACTIVITIES FOR HOME OR SCHOOL

Shell Study

Why is a spiral shell larger at one end?

Materials
- safety goggles
- gloves
- spiral shell from an animal such as a whelk, conch, or sea snail
- coarse sandpaper
- hand lens

Procedure

1. **CAUTION** Put on the safety goggles and gloves. Observe the outside of the shell. Rub the tip of the shell with sandpaper until you have a hole about 5 millimeters (about $\frac{1}{4}$ in.) wide.
2. Use the hand lens to observe the inside of the shell. What do you see?

Draw Conclusions

Animals that live in spiral shells usually keep their shells for their whole lives. When they begin their lives, they are very small. Their shells are very small too.

Think about the shell you observed. Why do you think the spiral is small at one end and gets bigger at the other end?

Feather Study

What are the parts of feathers?

Materials
- 1 or 2 types of feathers from a bird
- hand lens

Procedure

1. Study the feathers. Use the hand lens to look at their parts. Record what you observe.
2. Touch the feathers as you look at them. Record what you feel.

Draw Conclusions

Discuss what you know about birds. Think of ways the feathers help birds fly.

Chapter 2 Review and Test Preparation

Vocabulary Review

Use the terms below to complete the sentences 1 through 9. The page numbers in ( ) tell you where to look in the chapter if you need help.

inherit (A38)   **gills** (A51)
traits (A38)   **fish** (A52)
mammals (A42)   **scales** (A52)
birds (A45)   **reptiles** (A55)
amphibians (A50)

1. One trait of ____ is a body covering called fur.
2. The features a young animal gets from its parents are called ____.
3. Animals ____ their traits from their parents.
4. Animals that begin life in water and later live on land are ____.
5. ____ have bodies covered with feathers.

Mark each statement *True* or *False*. If a statement is false, change the underlined term to make the statement true.

6. Reptiles and fish have <u>scales</u>.
7. Scales make it easy for <u>mammals</u> to glide through the water.
8. Young amphibians and fish use <u>lungs</u> to take in oxygen from water.
9. <u>Amphibians</u> have dry, scaly skin.

Connect Concepts

Write the terms that belong in the concept map.

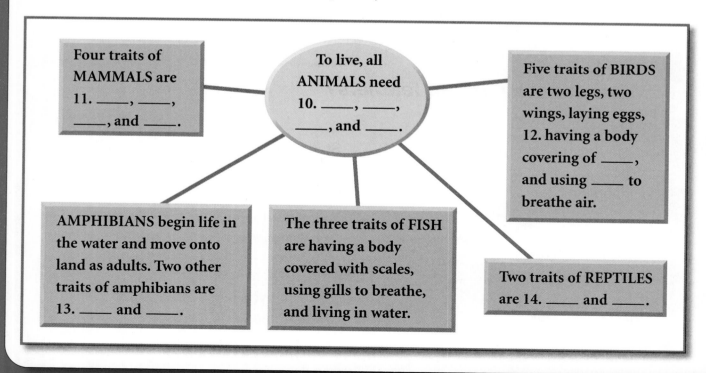

Four traits of MAMMALS are 11. ____, ____, ____, and ____.

To live, all ANIMALS need 10. ____, ____, ____, and ____.

Five traits of BIRDS are two legs, two wings, laying eggs, 12. having a body covering of ____, and using ____ to breathe air.

AMPHIBIANS begin life in the water and move onto land as adults. Two other traits of amphibians are 13. ____ and ____.

The three traits of FISH are having a body covered with scales, using gills to breathe, and living in water.

Two traits of REPTILES are 14. ____ and ____.

Check Understanding
Write the letter of the best choice.

15. What are two things animals use to get the air they need?
    A eyes and ears
    B scales and feathers
    C skin and fur
    D gills and lungs

16. What kind of shelter do beavers build?
    F dam   H den
    G lodge   J nest

17. Which kind of animal feeds its young with milk from its body?
    A fish   C amphibian
    B mammal   D bird

18. Young reptiles —
    F can meet their needs as soon as they are born
    G can swim
    H are carried in a pouch
    J begin life in the water

19. Which kind of animal is a salamander?
    A reptile   C fish
    B mammal   D amphibian

Critical Thinking

20. Describe how the body parts of a shark help it survive.

21. How are polar bears and bats alike?

Process Skills Review

22. Use what you have learned about animals to **classify** these animals into groups: snake, robin, dog, lizard, cow, goose, whale. Explain why you grouped the animals as you did.

23. Explain how making and **using a model** of a bird's nest can help you learn about birds.

24. How can you use your senses to **observe** birds?

Performance Assessment
Grouping Animals

Use the animal cards from the Investigate in Lesson 1. Group them by using what you have learned about animals. Are there any animals that don't belong in a group? What traits did these animals have?

UNIT A
Unit Project Wrap Up

Here are some ideas for ways to wrap up your unit project.

Take Photographs
Use a camera to take photographs to illustrate your handbook. Label each photograph with the time and date it was taken, and identify the subject.

Draw a Map
Draw a map of the area in which you made your observations. Show features like streams and forests. Draw on the map the plants and animals you found in the places you found them. See what conclusions you can draw about the needs of plants and animals based on where they live.

Make a Graph
Find a way to sort the living and nonliving things you found. Then make a bar graph to organize your data. You may want to use a computer graphing program.

Investigate Further
How could you make your project better? What other questions do you have about plants and animals? Plan ways to find answers to your questions. Use the Science Handbook on pages R2-R9 for help.

References

Science Handbook
Planning an Investigation	**R2**
Using Science Tools	**R4**
Using a Hand Lens	R4
Using a Thermometer	R4
Caring for and Using a Microscope	R5
Using a Balance	R6
Using a Spring Scale	R6
Measuring Liquids	R7
Using a Ruler or Meterstick	R7
Using a Timing Device	R7
Using a Computer	R8
Table of Measurements	R10

Health Handbook — R11

Glossary — R46

Index — R54

Planning an Investigation

When scientists observe something they want to study, they use scientific inquiry to plan and conduct their study. They use science process skills as tools to help them gather, organize, analyze, and present their information. This plan will help you work like a scientist.

Which food does my hamster eat the most of?

Step 1—Observe and ask questions.

- Use your senses to make observations.
- Record a question you would like to answer.

My hypothesis: My hamster will eat more sunflower seeds than any other food.

Step 2—Make a hypothesis.

- Choose one possible answer, or hypothesis, to your question.
- Write your hypothesis in a complete sentence.
- Think about what investigation you can do to test your hypothesis.

I'll give my hamster equal amounts of three kinds of foods, then observe what she eats.

Step 3—Plan your test.

- Write down the steps you will follow to do your test. Decide how to conduct a fair test by controlling variables.
- Decide what equipment you will need.
- Decide how you will gather and record your data.

Step 4—Conduct your test.

I'll repeat this experiment for four days. I'll meaure how much food is left each time.

- Follow the steps you wrote.
- Observe and measure carefully.
- Record everything that happens.
- Organize your data so that you can study it carefully.

Step 5—Draw conclusions and share results.

My hypothesis was correct. She ate more sunflower seeds than the other kinds of foods.

- Analyze the data you gathered.
- Make charts, graphs, or tables to show your data.
- Write a conclusion. Describe the evidence you used to determine whether your test supported your hypothesis.
- Decide whether your hypothesis was correct.

Investigate Further

I wonder if there are other foods she will eat . . .

Using Science Tools

Using a Hand Lens

1. Hold the hand lens about 12 centimeters (5 in.) from your eye.
2. Bring the object toward you until it comes into focus.

Using a Thermometer

1. Place the thermometer in the liquid. Never stir the liquid with the thermometer. Don't touch the thermometer any more than you need to. If you are measuring the temperature of the air, make sure that the thermometer is not in line with a direct light source.
2. Move so that your eyes are even with the liquid in the thermometer.
3. If you are measuring a material that is not being heated or cooled, wait about two minutes for the reading to become stable, or stay the same. Find the scale line that meets the top of the liquid in the thermometer, and read the temperature.
4. If the material you are measuring is being heated or cooled, you will not be able to wait before taking your measurements. Measure as quickly as you can.

Caring for and Using a Microscope

Caring for a Microscope

- Carry a microscope with two hands.
- Never touch any of the lenses of a microscope with your fingers.

Using a Microscope

1. Raise the eyepiece as far as you can using the coarse-adjustment knob. Place your slide on the stage.
2. Start by using the lowest power. The lowest-power lens is usually the shortest. Place the lens in the lowest position it can go to without touching the slide.
3. Look through the eyepiece, and begin adjusting it upward with the coarse-adjustment knob. When the slide is close to being in focus, use the fine-adjustment knob.
4. When you want to use a higher-power lens, first focus the slide under low power. Then, watching carefully to make sure that the lens will not hit the slide, turn the higher-power lens into place. Use only the fine-adjustment knob when looking through the higher-power lens.

You may use a Brock microscope. This sturdy microscope has only one lens.

1. Place the object to be viewed on the stage.
2. Look through the eyepiece, and raise the tube until the object comes into focus.

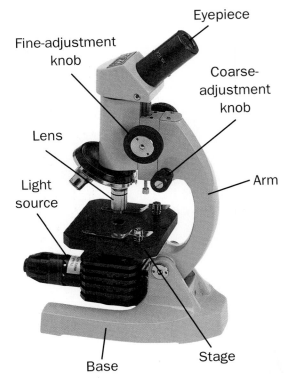

A Light Microscope

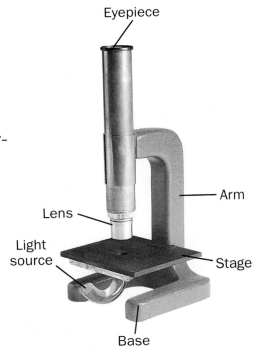

A Brock Microscope

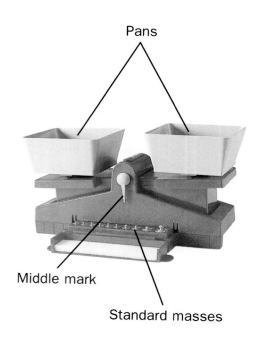

Pans
Middle mark
Standard masses

Using a Balance

1. Look at the pointer on the base to make sure the empty pans are balanced. Place the object you wish to measure in the left-hand pan.
2. Add the standard masses to the other pan. As you add masses, you should see the pointer move. When the pointer is at the middle mark, the pans are balanced.
3. Add the numbers on the masses you used. The total is the mass in grams of the object you measured.

Using a Spring Scale

Measuring an Object at Rest

1. Hook the spring scale to the object.
2. Lift the scale and object with a smooth motion. Do not jerk them upward.
3. Wait until any motion of the spring comes to a stop. Then read the number of newtons from the scale.

Measuring an Object in Motion

1. With the object resting on a table, hook the spring scale to it.
2. Pull the object smoothly across the table. Do not jerk the object.
3. As you pull, read the number of newtons you are using to pull the object.

Measuring Liquids

1. Pour the liquid you want to measure into a measuring container. Put your measuring container on a flat surface, with the measuring scale facing you.
2. Look at the liquid through the container. Move so that your eyes are even with the surface of the liquid in the container.
3. To read the volume of the liquid, find the scale line that is even with the surface of the liquid.
4. If the surface of the liquid is not exactly even with a line, estimate the volume of the liquid. Decide which line the liquid is closer to, and use that number.

Beaker **Graduate**

Using a Ruler or Meterstick

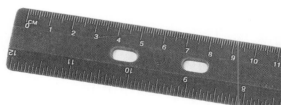

1. Place the zero mark or end of the ruler or meterstick next to one end of the distance or object you want to measure.
2. On the ruler or meterstick, find the place next to the other end of the distance or object.
3. Look at the scale on the ruler or meterstick. This will show the distance or the length of the object.

Using a Timing Device

1. Reset the stopwatch to zero.
2. When you are ready to begin timing, press *Start*.
3. As soon as you are ready to stop timing, press *Stop*.
4. The numbers on the dial or display show how many minutes, seconds, and parts of seconds have passed.

Using a Computer

Writing Reports

To write a report with a computer, use a word processing software program. After you are in the program, type your report. By using certain keys and the mouse, you can control how the words look, move words, delete or add words and copy them, check your spelling, and print your report.

Save your work to the desktop or hard disk of the computer, or to a floppy disk. You can go back to your saved work later if you want to revise it.

There are many reasons for revising your work. You may find new information to add or mistakes you want to correct. You may want to change the way you report your information because of who will read it.

Computers make revising easy. You delete what you don't want, add the new parts, and then save. You can also save different versions of your work.

For a science lab report, it is important to show the same kinds of information each time. With a computer, you can make a general format for a lab report, save the format, and then use it again and again.

Making Graphs and Charts

You can make a graph or chart with most word processing software programs. You can also use special software programs such as Data ToolKit or Graph Links. With Graph Links you can make pictographs and circle, bar, line, and double-line graphs.

SCIENCE HANDBOOK

First, decide what kind of graph or chart will best communicate your data. Sometimes it's easiest to do this by sketching your ideas on paper. Then you can decide what format and categories you need for your graph or chart. Choose that format for the program. Then type your information. Most software programs include a tutor that gives you step-by-step directions for making a graph or chart.

Doing Research

Computers can help you find current information from all over the world through the Internet. The Internet connects thousands of computer sites that have been set up by schools, libraries, museums, and many other organizations.

Get permission from an adult before you log on to the Internet. Find out the rules for Internet use at school or at home. Then log on and go to a search engine, which will help you find what you need. Type in keywords, words that tell the subject of your search. If you get too much information that isn't exactly about the topic, make your keywords more specific. When you find the information you need, save it or print it.

Harcourt Science tells you about many Internet sites related to what you are studying. To find out about these sites, called Web sites, look for Technology Links in the lessons in this book.

If you need to contact other people to help in your research, you can use e-mail. Log into your e-mail program, type the address of the person you want to reach, type your message, and send it. Be sure to have adult permission before sending or receiving e-mail.

Another way to use a computer for research is to access CD-ROMs. These are discs that look like music CDs. CD-ROMs can hold huge amounts of data, including words, still pictures, audio, and video. Encyclopedias, dictionaries, almanacs, and other sources of information are available on CD-ROMs. These computer discs are valuable resources for your research.

Measurement Systems

SI Measures (Metric)

Temperature
Ice melts at 0 degrees Celsius (°C)
Water freezes at 0°C
Water boils at 100°C

Length and Distance
1000 meters (m) = 1 kilometer (km)
100 centimeters (cm) = 1 m
10 millimeters (mm) = 1 cm

Force
1 newton (N) = 1 kilogram ×
  meter/second/second (kg-m/s^2)

Volume
1 cubic meter (m^3) = 1m × 1m × 1m
1 cubic centimeter (cm^3) =
  1 cm × 1 cm × 1 cm
1 liter (L) = 1000 milliliters (mL)
1 cm^3 = 1 mL

Area
1 square kilometer (km^2) =
  1 km × 1 km
1 hectare = 10 000 m^2

Mass
1000 grams (g) = 1 kilogram (kg)
1000 milligrams (mg) = 1 g

Rates (Metric and Customary)
kmh = kilometers per hour
m/s = meters per second
mph = miles per hour

Customary Measures

Volume of Fluids
8 fluid ounces (fl oz) = 1 cup (c)
2 c = 1 pint (pt)
2 pt = 1 quart (qt)
4 qt = 1 gallon (gal)

Temperature
Ice melts at 32 degrees
  Fahrenheit (°F)
Water freezes at 32°F
Water boils at 212°F

Length and Distance
12 inches (in) = 1 foot (ft)
3 ft = 1 yard (yd)
5,280 ft = 1 mile (mi)

Weight
16 ounces (oz) = 1 pound (lb)
2,000 pounds = 1 ton (T)

Health Handbook

Safety and First Aid
A Safe Bike	R12
Your Bike Helmet	R13
Fire Safety	R14
Earthquake Safety	R15
First Aid for Choking and Bleeding	R16
First Aid for Nosebleeds and Insect Bites	R18
When Home Alone	R20

Being Physically Active
Planning Your Weekly Activities	R22
Guidelines for a Good Workout	R23
Warm-Up and Cool-Down Stretches	R24
Building a Strong Heart and Lungs	R26

Nutrition and Food Safety
The Food Guide Pyramid	R28
Estimating Serving Sizes	R29
Fight Bacteria	R30
Food Safety Tips	R31

Caring for Your Body Systems
Sense Organs	R32
Skeletal System	R34
Muscular System	R36
Digestive System	R38
Circulatory System	R40
Respiratory System	R42
Nervous System	R44

Bicycle Safety

A Safe Bike

You probably know how to ride a bike, but do you know how to make your bike as safe as possible? A safe bike is the right size for you. When you sit on your bike with the pedal in the lowest position, you should be able to rest your heel on the pedal. Your body should be 2 inches (about 5 cm) above the support bar that goes from the handlebar stem to the seat support when you are standing astride your bike with both feet flat on the ground. After checking for the right size, check your bike for the safety equipment shown below. How safe is *your* bike?

HEALTH HANDBOOK

Your Bike Helmet

About 400,000 children are involved in bike-related crashes every year. That's why it's important to *always* wear your bike helmet. Wear your helmet flat on your head. Be sure it is strapped snugly so that the helmet will stay in place if you fall. If you do fall and strike your helmet on the ground, replace it, even if it doesn't look damaged. The padding inside the helmet may be crushed, which reduces the ability of the helmet to protect your head in the event of another fall. Look for the features shown here when purchasing a helmet.

- approval sticker
- quick-release strap
- padding
- hard shell
- air vent

Safety on the Road

Here are some tips for safe bicycle riding.

- Check your bike every time you ride it. Is it in safe working condition?
- Ride in single file in the same direction as traffic. Never weave in and out of parked cars.
- Before you enter a street, **STOP. Look** left, then right, then left again. **Listen** for any traffic. **Think** before you go.
- Walk your bike across an intersection. **Look** left, then right, then left again. Wait for traffic to pass.
- Obey all traffic signs and signals.
- Do not ride your bike at night without an adult. Be sure to wear light-colored clothing and use reflectors and front and rear lights for night riding.

Safety in Emergencies

Fire Safety

Fires cause more deaths than any other type of disaster. But a fire doesn't have to be deadly if you prepare your home and follow some basic safety rules.

- Install smoke detectors outside sleeping areas and on every other floor of your home. Test the detectors once a month and change the batteries twice a year.
- Keep a fire extinguisher on each floor of your home. Check them monthly to make sure they are properly charged.
- Make a fire escape plan. Ideally, there should be two routes out of each room. Sleeping areas are most important, as most fires happen at night. Plan to use stairs only, as elevators can be dangerous in a fire.
- Pick a place outside for everyone to meet. Choose one person to go to a neighbor's home to call 911 or the fire department.
- Practice crawling low to avoid smoke.
- If your clothes catch fire, follow the three steps shown here.

1. STOP
2. DROP
3. ROLL

Earthquake Safety

An earthquake is a strong shaking or sliding of the ground. The tips below can help you and your family stay safe in an earthquake.

Before an Earthquake	During an Earthquake	After an Earthquake
• Attach tall, heavy furniture, such as bookcases, to the wall. Store the heaviest items on the lowest shelves. • Check for fire risks. Bolt down gas appliances, and use flexible hosing and connections for both gas and water lines. • Strengthen and anchor overhead light fixtures to help keep them from falling.	• If you are outdoors, stay outdoors and move away from buildings and utility wires. • If you are indoors, take cover under a heavy desk or table, or in a doorway. Stay away from glass doors and windows and from heavy objects that might fall. • If you are in a car, drive to an open area away from buildings and overpasses.	• Keep watching for falling objects as aftershocks shake the area. • Check for hidden structural problems. • Check for broken gas, electric, and water lines. If you smell gas, shut off the gas main. Leave the area. Report the leak.

Storm Safety

- **In a Tornado** Take cover in a sheltered area away from doors and windows. An interior hallway or basement is best. Stay in the shelter until the danger has passed.
- **In a Hurricane** Prepare for high winds by securing objects outside or bringing them indoors. Cover windows and glass with plywood. Listen to weather bulletins for instructions. If asked to evacuate, proceed to emergency shelters.
- **In a Winter Storm or Blizzard** Stock up on food that does not have to be cooked. Dress in thin layers that help trap the body's heat. Pay special attention to the head and neck. If you are caught in a vehicle, turn on the dome light to make the vehicle visible to search crews.

First Aid

For Choking . . .

The tips on the next few pages can help you provide simple first aid to others and yourself. Always tell an adult about any injuries that occur.

If someone else is choking . . .

  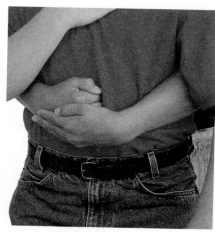

1. Recognize the Universal Choking Sign—grasping the throat with both hands. This sign means a person is choking and needs help.

2. Put your arms around his or her waist. Make a fist and put it above the person's navel. Grab your fist with your other hand.

3. Pull your hands toward yourself and give five quick, hard, upward thrusts on the choker's belly.

If you are choking when alone . . .

 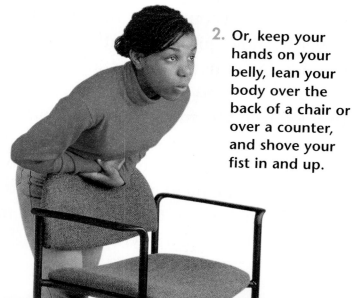

1. Make a fist and place it above your navel. Grab your fist with your other hand. Pull your hands up with a quick, hard thrust.

2. Or, keep your hands on your belly, lean your body over the back of a chair or over a counter, and shove your fist in and up.

For Bleeding . . .

If someone else is bleeding . . .

Wash your hands with soap, if possible.

Put on protective gloves, if available.

Wash small wounds with soap and water. Do *not* wash serious wounds.

Place a clean gauze pad or cloth over the wound. Press firmly for ten minutes. Don't lift the gauze during this time.

If you don't have gloves, have the injured person hold the cloth in place with his or her own hand.

If after ten minutes the bleeding has stopped, bandage the wound. If the bleeding has not stopped, continue pressing on the wound and get help.

If you are bleeding . . .

- Follow the steps shown above. You don't need gloves to touch your own blood.
- Be sure to tell an adult about your injury.

First Aid

For Nosebleeds . . .

- Sit down, and tilt your head forward. Pinch your nostrils together for at least ten minutes.
- You can also put an ice pack on the bridge of your nose.
- If your nose continues to bleed, get help from an adult.

For Burns . . .

Minor burns are called first degree burns and involve only the top layer of skin. The skin is red and dry and the burn is painful. More serious burns are called second or third degree burns. These burns involve the top and lower layers of skin. Second degree burns cause blisters, redness, swelling, and pain. Third degree burns are the most serious. The skin is gray or white and looks burned. All burns need immediate first aid.

Minor Burns

- Run cool water over the burn or soak it in cool water for at least five minutes.
- Cover the burn with a clean, dry bandage.
- Do *not* put lotion or ointment on the burn.

More Serious Burns

- Cover the burn with a cool, wet bandage or cloth. Do *not* break any blisters.
- Do *not* put lotion or ointment on the burn.
- Get help from an adult right away.

HEALTH HANDBOOK

For Insect Bites and Stings . . .

▲ deer tick

- Always tell an adult about bites and stings.
- Scrape out the stinger with your fingernail.
- Wash the area with soap and water.
- Ice cubes will usually take away the pain from insect bites. A paste made from baking soda and water also helps.
- If the bite or sting is more serious and is on the arm or leg, keep the leg or arm dangling down. Apply a cold, wet cloth. Get help immediately!
- If you find a tick on your skin, remove it. Crush it between two rocks. Wash your hands right away.
- If a tick has already bitten you, do not pull it off. Cover it with oil and wait for it to let go, then remove it with tweezers. Wash the area and your hands.

For Skin Rashes from Plants . . .

▲ poison ivy

Many poisonous plants have three leaves. Remember, "Leaves of three, let them be." If you touch a poisonous plant, wash the area. Put on clean clothes and throw the dirty ones in the washer. If a rash develops, follow these tips.

- Apply calamine lotion or a baking soda and water paste. Try not to scratch. Tell an adult.
- If you get blisters, do *not* pop them. If they burst, keep the area clean and dry. Cover with a bandage.
- If your rash does not go away in two weeks or if the rash is on your face or in your eyes, see your doctor.

Being Safe at Home

When Home Alone

Everyone stays home alone sometimes. When you stay home alone, it's important to know how to take care of yourself. Here are some easy rules to follow that will help keep you safe when you are at home by yourself.

Do These Things

- Lock all the doors and windows. Be sure you know how to lock and unlock all the locks.

- If someone calls who is nasty or mean, hang up. Your parents may not want you to answer the phone at all.

- If you have an emergency, call 911 or 0 (zero) for the operator. Describe the problem, give your full name, address, and telephone number. Follow all instructions given to you.

- If you see anyone hanging around outside, tell an adult or call the police.

- If you see or smell smoke, go outside right away. If you live in an apartment, do not take the elevator. Go to a neighbor's home and call 911 or the fire department immediately.

- Entertain yourself. Time will pass more quickly if you are not bored. Try not to spend your time watching television. Instead, work on a hobby, read a book or magazine, do your homework, or clean your room. Before you know it, an adult will be home.

Do NOT Do These Things

- Do NOT use the stove, microwave, or oven unless an adult family member has given you permission, and you are sure about how to use these appliances.

- Do NOT open the door for anyone you don't know or for anyone who is not supposed to be in your home.
  - If someone rings the bell and asks to use the telephone, tell the person to go to a phone booth.
  - If someone tries to deliver a package, do NOT open the door. The delivery person will leave the package or come back later.
  - If someone is selling something, do NOT open the door. Just say, "We're not interested," and nothing more.

- Do NOT talk to strangers on the telephone. Do not tell anyone that you are home alone. If the call is for an adult family member, say that they can't come to the phone right now and take a message. Ask for the caller's name and phone number and deliver the message when an adult family member comes home.

- Do NOT have friends over unless you have gotten permission from your parents or other adult family members.

Being Physically Active

Planning Your Weekly Activities

Being active every day is important for your overall health. Physical activity helps you manage stress, maintain a healthful weight, and strengthen your body systems. The Activity Pyramid, like the Food Guide Pyramid, can help you choose a variety of activities in the right amounts to keep your body strong and healthy.

The Activity Pyramid

Sitting for more than thirty minutes at a time: Only Once in a While

Light Exercise: Two to Three Times a Week

Flexibility and Strength: Two to Three Times a Week

Twenty-plus minutes of continuous aerobic activity: Three to Five Times a Week

Stay active: Every Day

Guidelines for a Good Workout

There are three things you should do every time you are going to exercise—warm up, work out, and cool down.

Warm-Up When you warm up, your heart rate, breathing rate, and body temperature increase and more blood flows to your muscles. As your body warms up, you can move more easily. People who warm up are less stiff after exercising, and are less likely to have exercise-related injuries. Your warm-up should include five minutes of stretching, and five minutes of low-level exercise.

Workout The main part of your exercise routine should be an aerobic exercise that lasts 20 to 30 minutes. Aerobic exercises make your heart, lungs, and circulatory system stronger.

Some common aerobic exercises are shown on pages R26–R27. You may want to mix up the types of activities you do. This helps you work different muscles, and provides a better workout over time.

Cool-Down When you finish your aerobic exercise, you need to give your body time to cool down. Start your cool-down with three to five minutes of low-level activity. End with stretching exercises to prevent soreness and stiffness.

Being Physically Active

Warm-Up and Cool-Down Stretches

Before you exercise, you should warm up your muscles. The warm-up exercises shown here should be held for at least fifteen to twenty seconds and repeated at least three times.

At the end of your workout, spend about two minutes repeating some of these stretches.

▶ **Hurdler's Stretch**
HINT—Keep the toes of your extended leg pointed up.

▲ **Shoulder and Chest Stretch** HINT—Pulling your hands slowly toward the floor gives a better stretch. Keep your elbows straight, but not locked!

◀ **Sit-and-Reach Stretch**
HINT—Remember to bend at the waist. Keep your eyes on your toes!

HEALTH HANDBOOK

▲ **Upper Back and Shoulder Stretch**
HINT—Try to stretch your hand down so that it rests flat against your back.

▼ **Thigh Stretch** HINT—Keep both hands flat on the ground. Lean as far forward as you can.

▲ **Calf Stretch** HINT—Keep both feet on the floor during this stretch. Try changing the distance between your feet. Is the stretch better for you when your legs are closer together or farther apart?

Tips for Stretching

- Never bounce when stretching.
- Hold each stretch for fifteen to twenty seconds.
- Breathe normally. This helps your body get the oxygen it needs.
- Do NOT stretch until it hurts. Stretch only until you feel a slight pull.

Being Physically Active

Building a Strong Heart and Lungs

Aerobic activities cause deep breathing and a fast heart rate for at least twenty minutes. These activities help both your heart and your lungs. Because your heart is a muscle, it gets stronger with exercise. A strong heart doesn't have to work as hard to pump blood to the rest of your body. Exercise also allows your lungs to hold more air. With a strong heart and lungs, your cells get oxygen faster and your body works more efficiently.

▲ **Swimming** Swimming is great for your endurance and flexibility. Even if you're not a great swimmer, you can use a kickboard and have a great time and a great workout just kicking around the pool. Be sure to swim only when a lifeguard is present.

◀ **In-line Skating** Remember to always wear a helmet when skating. Always wear protective pads on your elbows and knees, and guards on your wrists, too. Learning how to skate, stop, and fall correctly will make you a safer skater.

HEALTH HANDBOOK

▲ **Jumping Rope** Jumping rope is one of the best ways to increase your endurance. Remember to always jump on an even surface and always wear supportive shoes.

▼ **Bicycling** Bicycling provides good aerobic activity *and* a great way to see the outdoors. Be sure to learn and follow bicycle safety rules. And *always* remember to wear your helmet!

▼ **Walking** A fast-paced walk is a terrific way to build your endurance. The only equipment you need is supportive shoes. Walking with a friend can make this exercise a lot of fun.

R27

Good Nutrition

The Food Guide Pyramid

No one food or food group supplies everything your body needs for good health. That's why it's important to eat foods from all the food groups. The Food Guide Pyramid can help you choose healthful foods in the right amounts. By choosing more foods from the groups at the bottom of the pyramid and fewer foods from the group at the top, you will eat the foods that provide your body with energy to grow and develop.

Fats, oils, and sweets
Eat sparingly.

Meat, poultry, fish, dry beans, eggs, and nuts 2–3 servings

Milk, yogurt, and cheese 2–3 servings

Fruits 2–4 servings

Vegetables 3–5 servings

Breads, cereals, rice, and pasta 6–11 servings

Estimating Serving Sizes

Choosing a variety of foods is only half the story. You also need to choose the right amounts. The table below can help you estimate the number of servings you are eating of your favorite foods.

Food Group	Amount of Food in One Serving	Some Easy Ways to Estimate Serving Size
Bread, Cereal, Rice, Pasta Group	1 ounce ready-to-eat (dry) cereal	large handful of plain cereal or a small handful of cereal with raisins and nuts
	1 slice bread, $\frac{1}{2}$ bagel	
	$\frac{1}{2}$ cup cooked pasta, rice, or cereal	ice cream scoop
Vegetable Group	1 cup of raw, leafy vegetables	about the size of a fist
	$\frac{1}{2}$ cup other vegetables, cooked or raw, chopped	
	$\frac{3}{4}$ cup vegetable juice	
	$\frac{1}{2}$ cup tomato sauce	ice cream scoop
Fruit Group	medium apple, pear, or orange	a baseball
	$\frac{1}{2}$ large banana or one medium banana	
	$\frac{1}{2}$ cup chopped or cooked fruit	
	$\frac{3}{4}$ cup of fruit juice	
Milk, Yogurt, and Cheese Group	$1\frac{1}{2}$ ounces of natural cheese	two dominoes
	2 ounces of processed cheese	$1\frac{1}{2}$ slices of packaged cheese
	1 cup of milk or yogurt	
Meat, Poultry, Fish, Dry Beans, Eggs, and Nuts Group	3 ounces of lean meat, chicken, or fish	about the size of your palm
	2 tablespoons peanut butter	
	$\frac{1}{2}$ cup of cooked dry beans	
Fats, Oils, and Sweets Group	1 teaspoon of margarine or butter	about the size of the tip of your thumb

Preparing Foods Safely

Fight Bacteria

You probably already know to throw away food that smells bad or looks moldy. But food doesn't have to look or smell bad to make you ill. To keep your food safe and yourself from becoming ill, follow the steps outlined in the picture below. And remember—when in doubt, throw it out!

HEALTH HANDBOOK

Food Safety Tips

Tips for Preparing Food

- Wash hands in warm, soapy water before preparing food. It's also a good idea to wash hands after preparing each dish.
- Defrost meat in the microwave or the refrigerator.
- Keep raw meat, poultry, fish, and their juices away from other food.
- Wash cutting boards, knives, and countertops immediately after cutting up meat, poultry, or fish. Never use the same cutting board for meats and vegetables without washing the board first.

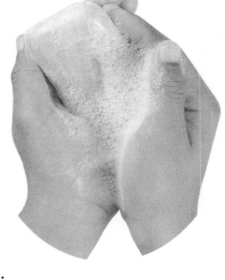

Tips for Cooking Food

- Cook all food completely, especially meat. Complete cooking kills the bacteria that can make you ill.
- Red meats should be cooked to a temperature of 160°F. Poultry should be cooked to 180°F. When done, fish flakes easily with a fork.
- Never eat food that contains raw eggs or raw egg yolks, including cookie dough.

Tips for Cleaning Up the Kitchen

- Wash all dishes, utensils, and countertops with hot, soapy water. Use a soap that kills bacteria, if possible.
- Store leftovers in small containers that will cool quickly in the refrigerator. Don't leave leftovers on the counter to cool.

Sense Organs

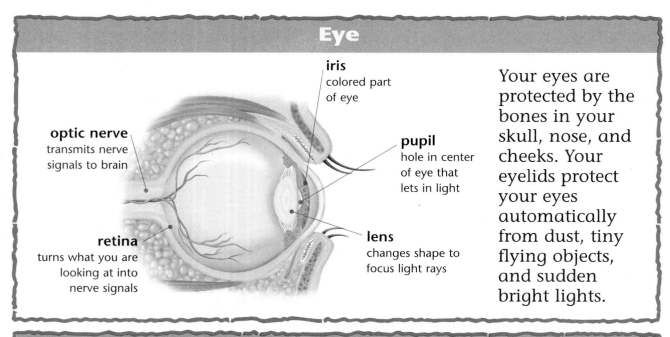

Eye

iris colored part of eye

optic nerve transmits nerve signals to brain

pupil hole in center of eye that lets in light

retina turns what you are looking at into nerve signals

lens changes shape to focus light rays

Your eyes are protected by the bones in your skull, nose, and cheeks. Your eyelids protect your eyes automatically from dust, tiny flying objects, and sudden bright lights.

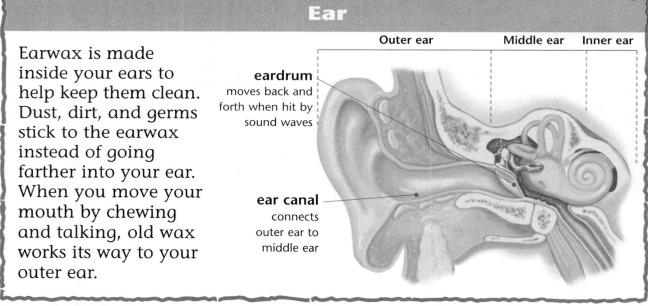

Ear

Outer ear   Middle ear   Inner ear

eardrum moves back and forth when hit by sound waves

ear canal connects outer ear to middle ear

Earwax is made inside your ears to help keep them clean. Dust, dirt, and germs stick to the earwax instead of going farther into your ear. When you move your mouth by chewing and talking, old wax works its way to your outer ear.

Caring for Your Eyes and Ears

- Have your eyesight (vision) checked every year.
- Wear safety glasses when participating in activities that can be dangerous to the eyes, such as sports and mowing grass.
- Wash in, around, and behind your outer ear. Do not try to clean your ear canal with cotton-tip sticks or other objects.

Nose

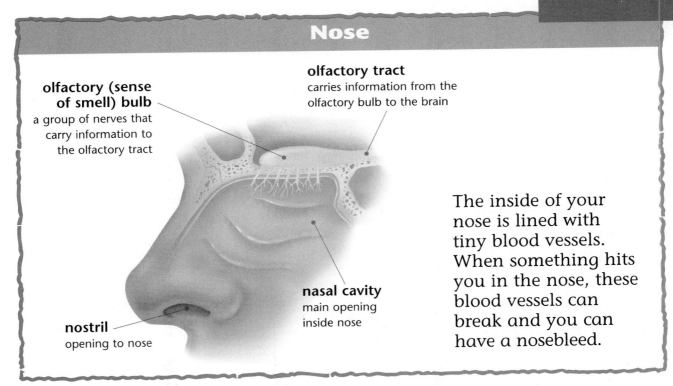

olfactory (sense of smell) bulb a group of nerves that carry information to the olfactory tract

olfactory tract carries information from the olfactory bulb to the brain

nostril opening to nose

nasal cavity main opening inside nose

The inside of your nose is lined with tiny blood vessels. When something hits you in the nose, these blood vessels can break and you can have a nosebleed.

Caring for Your Nose, Tongue, and Skin

- If you get a nosebleed, sit, lean forward slightly, and pinch just below the bridge of your nose for ten minutes. Breathe through your mouth.
- When you brush your teeth, brush your tongue too.
- Always wear sunscreen when you are in the sun.

Tongue

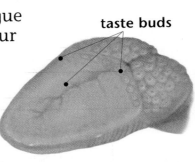

Germs live on your tongue and in other parts of your mouth. Germs can harm your teeth and give you bad breath.

taste buds

Skin

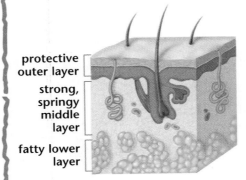

protective outer layer
strong, springy middle layer
fatty lower layer

Your skin protects your insides from the outside world. It keeps fluids you need inside your body and fluids you don't need, such as swimming pool water, outside your body.

Skeletal System

Each of your bones has a particular shape and size that allow it to do a certain job. You have bones that are tiny, long, wide, flat, and even curved. The job of some bones is to protect your body parts.

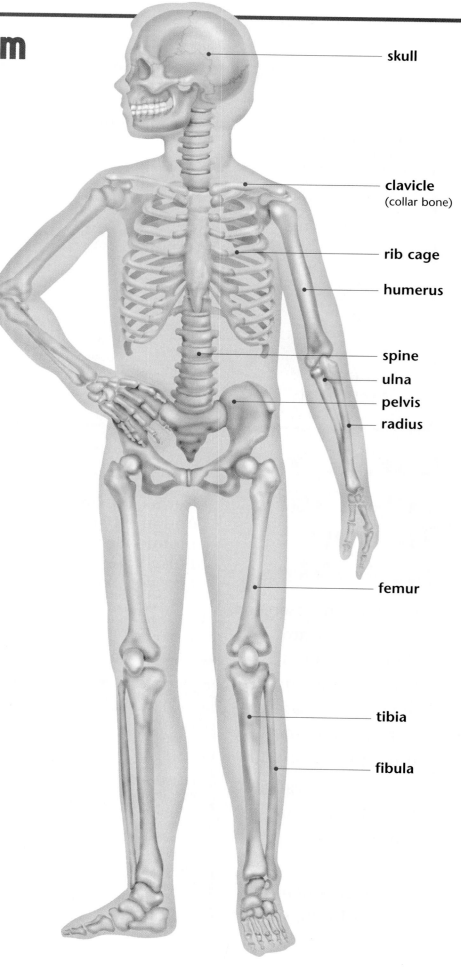

- skull
- clavicle (collar bone)
- rib cage
- humerus
- spine
- ulna
- pelvis
- radius
- femur
- tibia
- fibula

HEALTH HANDBOOK

Bones that Protect

Rib Cage Your rib bones form a cage that protects your heart and lungs from all sides. Your ribs are springy. When something strikes you in the chest, your ribs push the object away instead of letting it hit your heart and lungs.

Your ribs are connected to your breastbone (sternum) by springy material called cartilage. The springy connection lets your ribs move up and down. This happens when your rib cage gets bigger and smaller as you breathe in and out.

Skull The bones in your head are called your skull. Some of the bones in your skull protect your brain. The bones in your face are part of your skull too.

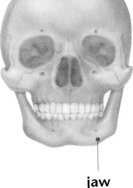

- cartilage
- rib
- sternum
- jaw

Caring for Your Skeletal System

- Calcium helps bones grow and makes them strong. Dairy products like milk, cheese, and yogurt contain calcium. Have 2–3 servings of dairy products every day.
- Exercise also makes your bones strong. When bones aren't used, they can become brittle and may break.

Activities

1. Look at the picture of the skeleton. Name a long bone. Name a short bone. Name a curved bone.

2. Put a tomato inside a wire cage. Gently throw a wad of paper at the cage. What happens? The cage protects the tomato in the same way your ribs protect your heart.

3. Measure around your rib cage with a string. How big is it when you breathe in? How big is it when you breathe out? Which measurement is greater?

R35

Muscular System

Like your bones, each muscle in your body does a certain job. Muscles in your thumb help you hold things. Muscles in your neck help you turn your head. Muscles in your arms help you pull or lift objects.

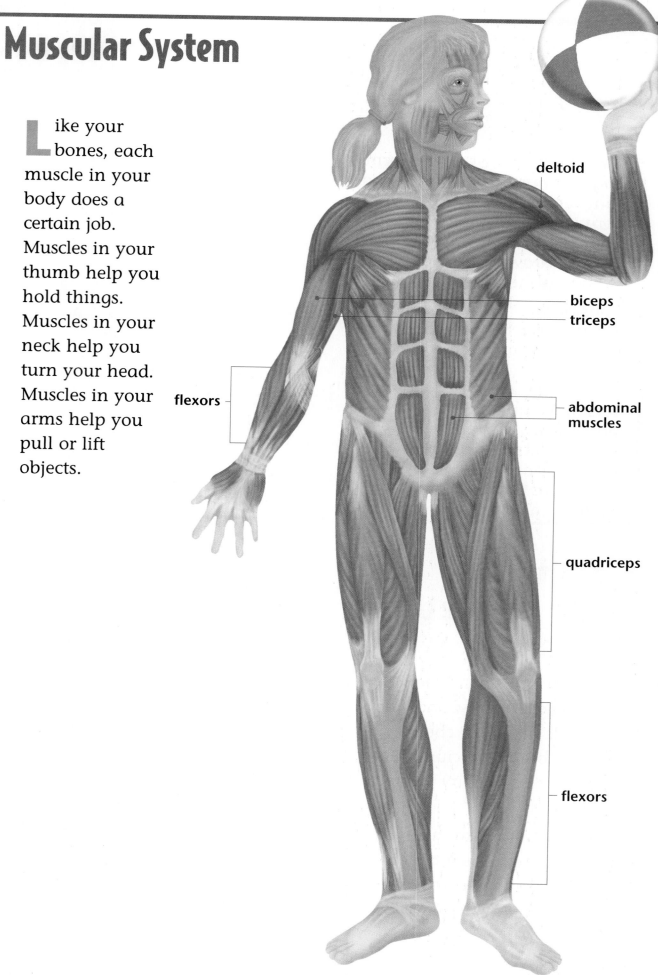

- deltoid
- biceps
- triceps
- flexors
- abdominal muscles
- quadriceps
- flexors

HEALTH HANDBOOK

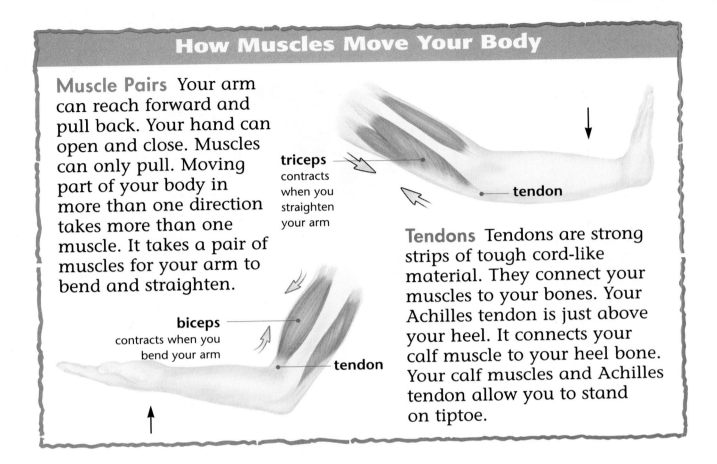

How Muscles Move Your Body

Muscle Pairs Your arm can reach forward and pull back. Your hand can open and close. Muscles can only pull. Moving part of your body in more than one direction takes more than one muscle. It takes a pair of muscles for your arm to bend and straighten.

triceps contracts when you straighten your arm

biceps contracts when you bend your arm

tendon

Tendons Tendons are strong strips of tough cord-like material. They connect your muscles to your bones. Your Achilles tendon is just above your heel. It connects your calf muscle to your heel bone. Your calf muscles and Achilles tendon allow you to stand on tiptoe.

Caring for Your Muscular System

- Exercise makes your muscles stronger.
- Stretching before you exercise makes muscles and tendons more flexible and less likely to get hurt.

Activities

1. **Tie your shoe without using your thumb. What happens?**
2. **Pull up on a desk with one hand. With your other hand, feel which arm muscle is working. Now push on the desk. Which arm muscle is working?**
3. **Ask a friend to push down on your arms for one minute while you push up as hard as you can. When your friend lets go, what happens?**

Digestive System

Food is broken down and pushed through your body by your digestive system. Your digestive system is a series of connected parts that starts with your mouth and ends with your large intestine.

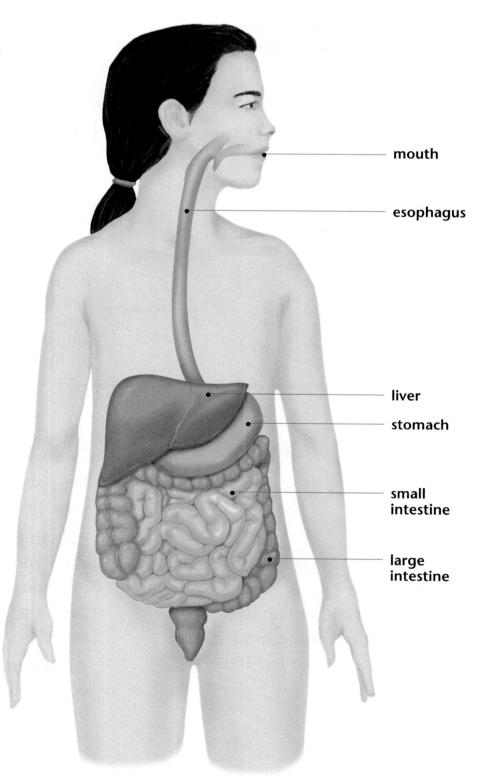

HEALTH HANDBOOK

From Mouth to Stomach

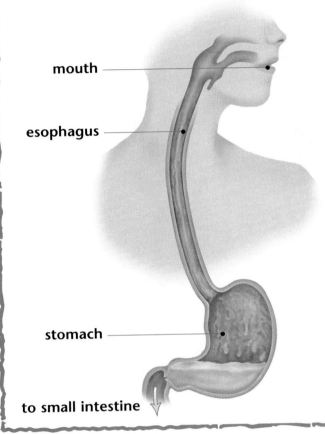

Esophagus Your esophagus, or food tube, is a tube that connects your mouth to your stomach. After you swallow a bite of food, muscles in your esophagus push the food into your stomach.

Stomach Your stomach is filled with acid that helps dissolve food. The stomach walls are strong muscles that mix food with the acid. The stomach walls are protected from the acid by a thick layer of mucus. From your stomach, food moves to the small intestine and then to the large intestine.

Caring for Your Digestive System

- Chew everything you eat carefully. Well-chewed food is easier to digest.
- Do not overeat. Overeating can cause a stomachache.

Activities

1. Measure 25 feet (about 8 m) on the floor. This is how long your digestive system is.
2. Cut a narrow balloon so that it is open on both ends. Put a wad of paper in one end. Squeeze the outside of the balloon to push the paper through and out the other end. This is similar to how your esophagus pushes food to your stomach.

Circulatory System

Food and oxygen travel through your circulatory system to every cell in your body. Blood moves nutrients throughout your body, fights infection, and helps control your body temperature. Your blood is made up mostly of a watery liquid called plasma.

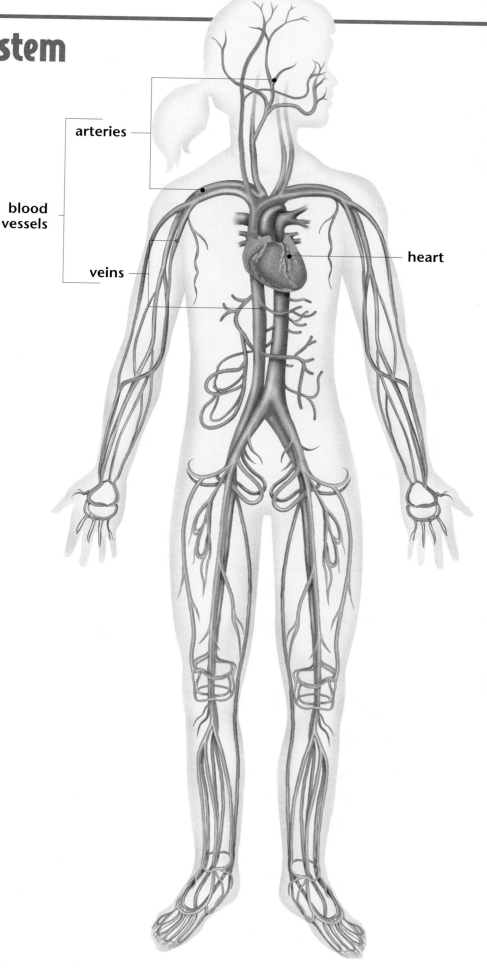

HEALTH HANDBOOK

Blood Vessels

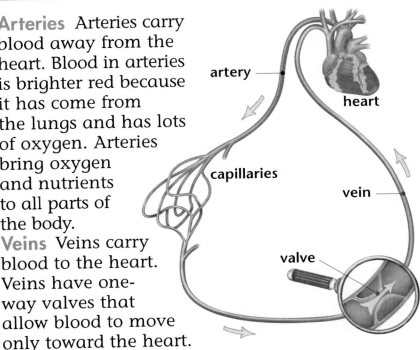

Arteries Arteries carry blood away from the heart. Blood in arteries is brighter red because it has come from the lungs and has lots of oxygen. Arteries bring oxygen and nutrients to all parts of the body.

Veins Veins carry blood to the heart. Veins have one-way valves that allow blood to move only toward the heart.

Capillaries Capillaries are very small pathways for blood. When blood flows through capillaries, it gives oxygen and nutrients to the cells in your body. Blood also picks up carbon dioxide and other waste.

Caring for Your Circulatory System

- Never touch another person's blood.
- Eat a healthy, balanced diet throughout your life to keep excess fat from blocking the blood flowing through your arteries.
- Get regular exercise to keep your heart strong.

Activities

1. Take the bottom out of a paper cup. Bend the top together like a clamshell. Hold the cup and drop a marble through from the bottom. Now try to drop one into the top. The clamshell-shaped cup is like the one-way valve in a vein.

2. Find the blue lines under the skin on your wrist. These are veins. Press gently and stroke along the lines toward your elbow. Now stroke toward your hand. What do you see?

R41

Respiratory System

Your body uses its respiratory system to get oxygen from the air and get rid of excess carbon dioxide. Your respiratory system is made up of your nose and mouth, your trachea (windpipe), your two lungs, and your diaphragm—a dome-shaped muscle under your lungs.

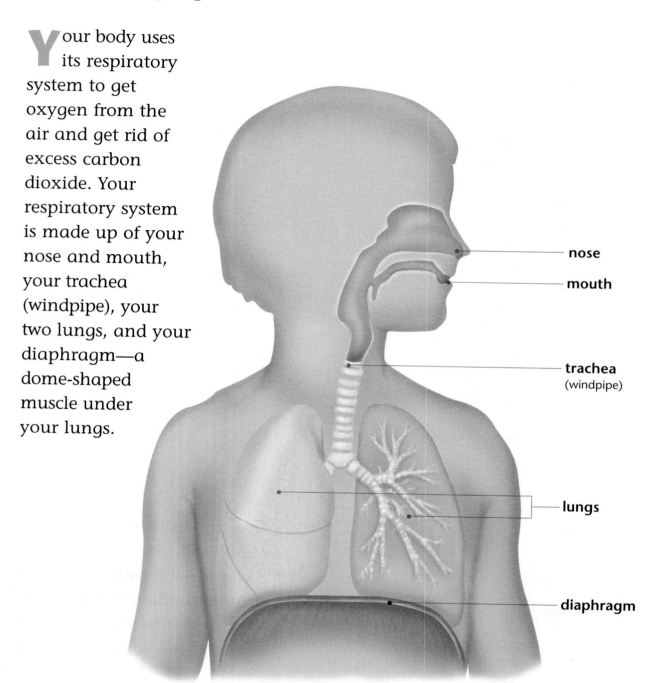

HEALTH HANDBOOK

Breathing

When you inhale, or breathe in, air enters your mouth and nose and goes into your trachea. Your trachea connects your nose and mouth to your lungs. Your trachea divides into two smaller tubes that go to your lungs. Your lungs fill with air. When you exhale, or breathe out, your diaphragm pushes upward. Air is forced up your trachea and out your mouth and nose.

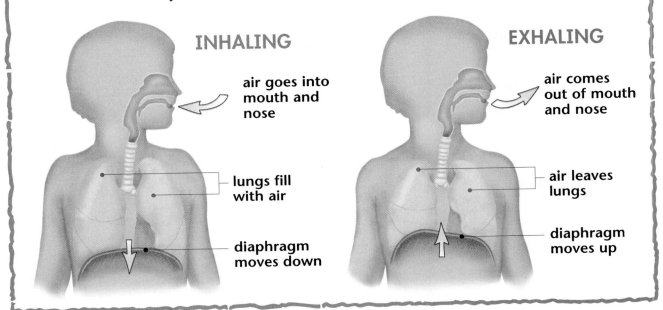

INHALING
- air goes into mouth and nose
- lungs fill with air
- diaphragm moves down

EXHALING
- air comes out of mouth and nose
- air leaves lungs
- diaphragm moves up

Caring for Your Respiratory System

- Exercise. When you exercise your body, you exercise your respiratory system too. Your muscles use more oxygen, so you breathe faster and deeper.
- Get enough sleep to help your resistance to colds.

Activities

1. Sit in a chair and count how many breaths you take in 30 seconds. Then exercise for two minutes. When you stop, count how many breaths you take in 30 seconds. Do you breathe more while sitting or after exercise?

2. Put your hand on your bellybutton and take a deep breath in and out. How does your hand move?

Nervous System

Your nerves send information to your brain from various parts of your body and from the outside world. Your brain decides what to do with the information and sends instructions through your nerves back to your body parts.

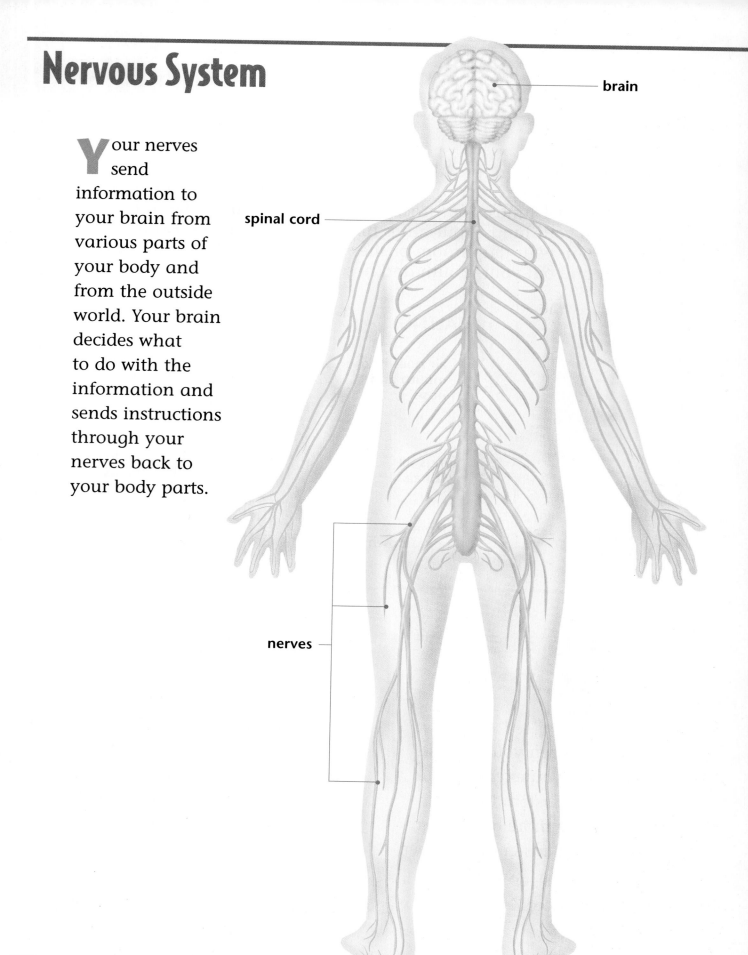

brain

spinal cord

nerves

HEALTH HANDBOOK

Your Brain

Your brain is about two pounds of wrinkled, pinkish-gray material. It's protected by your skull and cushioned by a thin layer of liquid. The brain's main connection to the body is the spinal cord.

Different parts of the brain send signals to different parts of your body. For example, the part right behind your forehead tells your body how to move. The area near the base of your neck controls your breathing and heartbeat. If you are left-handed, the right half of your brain controls your handwriting.

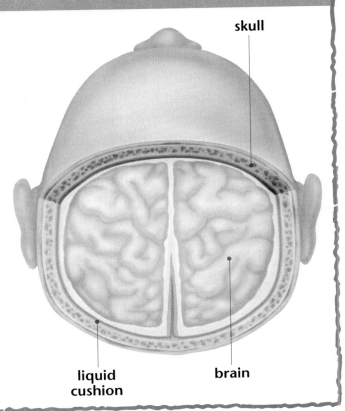

Caring for Your Nervous System

- Many injuries to the brain are caused by car crashes. Wear your safety belt and sit in the backseat when you are in the car.
- Always wear a helmet when you ride your bike, skate, or use a skateboard.

Activities

1. Make a list of signals your nerves are sending to your brain right now. Also list instructions your brain is sending to your nerves.
2. Read a paragraph out of a book while the television is on. Do you know what the paragraph was about? Do you know what happened on television?
3. Write your name with your opposite hand ten times. Does your writing improve?

Visit the Multimedia Science Glossary to see illustrations of these words and to hear them pronounced.
www.harcourtschool.com/scienceglossary

Glossary

This Glossary contains important science words and their definitions. Each word is respelled as it would be in a dictionary. When you see the ' mark after a syllable, pronounce that syllable with more force than the other syllables. The page number at the end of the definition tells where to find the word in your book. The boldfaced letters in the examples in the Pronunciation Key that follows show how these letters are pronounced in the respellings after each glossary word.

PRONUNCIATION KEY

a	**a**dd, m**a**p	m	**m**ove, see**m**	u	**u**p, d**o**ne
ā	**a**ce, r**a**te	n	**n**ice, ti**n**	û(r)	b**u**rn, t**er**m
â(r)	c**a**re, **ai**r	ng	ri**ng**, so**ng**	yōō	f**u**se, f**ew**
ä	p**a**lm, f**a**ther	o	**o**dd, h**o**t	v	**v**ain, e**v**e
b	**b**at, ru**b**	ō	**o**pen, s**o**	w	**w**in, a**w**ay
ch	**ch**eck, cat**ch**	ô	**o**rder, j**aw**	y	**y**et, **y**earn
d	**d**og, ro**d**	oi	**oi**l, b**oy**	z	**z**est, mu**s**e
e	**e**nd, p**e**t	ou	p**ou**t, n**ow**	zh	vi**si**on, plea**s**ure
ē	**e**qual, tr**ee**	ŏŏ	t**oo**k, f**u**ll	ə	the schwa, an unstressed vowel representing the sound spelled
f	**f**it, hal**f**	ōō	p**oo**l, f**oo**d		
g	**g**o, lo**g**	p	**p**it, sto**p**		
h	**h**ope, **h**ate	r	**r**un, poo**r**		
i	**i**t, g**i**ve	s	**s**ee, pa**ss**		*a* in **a**bove
ī	**i**ce, wr**i**te	sh	**sh**ure, ru**sh**		*e* in sick**e**n
j	**j**oy, le**dge**	t	**t**alk, si**t**		*i* in poss**i**ble
k	**c**ool, ta**k**e	th	**th**in, bo**th**		*o* in mel**o**n
l	**l**ook, ru**l**e	th	**th**is, ba**th**e		*u* in circ**u**s

Other symbols:
- separates words into syllables
- ' indicates heavier stress on a syllable
- ' indicates light stress on a syllable

Multimedia Science Glossary: **www.harcourtschool.com/scienceglossary**

GLOSSARY

absorption [ab•sôrp′shən] The stopping of light **(F40)**

amphibian [am•fib′ē•ən] An animal that begins life in the water and moves onto land as an adult **(A50)**

anemometer [an′ə•mom′ə•tər] An instrument that measures wind speed **(D40)**

asteroid [as′tər•oid] A chunk of rock that orbits the sun **(D64)**

atmosphere [at′məs•fir′] The air that surrounds Earth **(D30)**

atom [at′əm] The basic building block of matter **(E16)**

axis [ak′sis] An imaginary line that goes through the North Pole and the South Pole of Earth **(D68)**

barrier island [bar′ē•ər ī′lənd] A landform; a thin island along a coast **(C35)**

bird [bûrd] An animal that has feathers, two legs, and wings **(A45)**

canyon [kan′yən] A landform; a deep valley with very steep sides **(C35)**

chemical change [kem′i•kəl chānj′] A change that forms different kinds of matter **(E46)**

chlorophyll [klôr′ə•fil′] The substance that gives plants their green color; it helps a plant use energy from the sun to make food **(A20)**

clay [klā] A type of soil made up of very small grains; it holds water well **(C69)**

coastal forest [kōs′təl fôr′ist] A thick forest with tall trees that gets a lot of rain and does not get very warm or cold **(B15)**

comet [kom′it] A large ball of ice and dust that orbits the sun **(D64)**

community [kə•myōō′nə•tē] All the populations of organisms that live in an ecosystem **(B7)**

condensation [kon′dən•sā′shən] The changing of a gas into a liquid **(D17)**

conductor [kən•duk′tər] A material in which thermal energy moves easily **(F15)**

coniferous forest [kō•nif′ər•əs fôr′ist] A forest in which most of the trees are conifers (cone-bearing) and stay green all year **(B16)**

conservation [kon′ser•vā′shən] The saving of resources by using them carefully **(C76)**

R47

constellation [kon′stə•lā′shən] A group of stars that form a pattern **(D84)**

consumer [kən•sōōm′ər] A living thing that eats other living things as food **(B43)**

contour plowing [kon′tōōr plou′ing] A type of plowing for growing crops; creates rows of crops around the sides of a hill instead of up and down **(C76)**

core [kôr] The center of the Earth **(C8)**

crust [krust] The solid outside layer of the Earth **(C8)**

deciduous forest [dē•sij′ōō•əs fôr′ist] A forest in which most of the trees lose and regrow their leaves each year **(B13)**

decomposer [dē′kəm•pōz′er] A living thing that breaks down dead organisms for food **(B44)**

desert [dez′ərt] An ecosystem where there is very little rain **(B20)**

earthquake [ûrth′kwāk′] The shaking of Earth's surface caused by movement of the crust and mantle **(C48)**

ecosystem [ek′ō•sis′təm] The living and nonliving things in an environment **(B7)**

energy [en′ər•jē] The ability to cause change **(F6)**

energy pyramid [en′ər•jē pir′ə•mid] A diagram that shows that the amount of useable energy in an ecosystem is less for each higher animal in the food chain **(B50)**

environment [in•vī′rən•mənt] The things, both living and nonliving, that surround a living thing **(B6)**

erosion [i•rō′zhən] The movement of weathered rock and soil **(C42)**

estuary [es′chōō•er′•ē] A place where fresh water from a river mixes with salt water from the ocean **(D12)**

evaporation [ē•vap′ə•rā′shən] The process by which a liquid changes into a gas **(D17, E18)**

fish [fish] An animal that lives its whole life in water and breathes with gills **(A52)**

flood [flud] A large amount of water that covers normally dry land **(C50)**

food chain [fōōd′ chān′] The path of food from one living thing to another **(B48)**

food web [fōōd′ web′] A model that shows how food chains overlap **(B54)**

Multimedia Science Glossary: www.harcourtschool.com/scienceglossary

GLOSSARY

force [fôrs] A push or a pull **(F58)**

forest [fôr′ist] An area in which the main plants are trees **(B12)**

fossil [fos′əl] Something that has lasted from a living thing that died long ago **(C20)**

fresh water [fresh′ wôt′ər] Water that has very little salt in it **(B26)**

front [frunt] A place where two air masses of different temperatures meet **(D37)**

gas [gas] A form of matter that does not have a definite shape or a definite volume **(E12)**

germinate [jûr′mə•nāt′] When a new plant breaks out of the seed **(A13)**

gills [gilz] A body part found in fish and young amphibians that takes in oxygen from the water **(A51)**

glacier [glā′shər] A huge sheet of ice **(C44)**

gravity [grav′i•tē] The force that pulls objects toward each other **(F62)**

groundwater [ground′wôt′ər] A form of fresh water that is found under Earth's surface **(D8)**

habitat [hab′ə•tat′] The place where a population lives in an ecosystem **(B7)**

heat [hēt] The movement of thermal energy from one place to another **(F8)**

humus [hyoo′məs] The part of the soil made up of decayed parts of once-living things **(C62)**

igneous rock [ig′nē•əs rok′] A rock that was once melted rock but has cooled and hardened **(C12)**

inclined plane [in•klīnd′ plān′] A simple machine made of a flat surface set at an angle to another surface **(F71)**

inexhaustible resource [in′eg•zôs′tə•bəl rē′sôrs] A resource such as air or water that can be used over and over and can't be used up **(C94)**

inherit [in•her′it] To receive traits from parents **(A38)**

insulator [in′sə•lāt′ər] A material in which thermal energy does not move easily **(F15)**

interact [in′tər•akt′] When plants and animals affect one another or the environment to meet their needs **(B42)**

R49

landform [land′fôrm′] A natural shape or feature of Earth's surface **(C34)**

leaf [lēf] A plant part that grows out of the stem; it takes in the air and light that a plant needs **(A7)**

lever [lev′ər] A bar that moves on or around a fixed point **(F70)**

liquid [lik′wid] A form of matter that has volume that stays the same, but can change its shape **(E12)**

loam [lōm] A type of topsoil that is rich in minerals and has lots of humus **(C70)**

lunar eclipse [lōō′nər i•klips′] The hiding of the moon when it passes through the Earth's shadow **(D78)**

mammal [mam′əl] An animal that has fur or hair and is fed milk from its mother's body **(A42)**

mantle [man′təl] The middle layer of the Earth **(C8)**

mass [mas] The amount of matter in an object **(E24)**

matter [mat′ər] Anything that takes up space **(E6)**

metamorphic rock [met′ə•môr′fik rok′] A rock that has been changed by heat and pressure **(C12)**

mineral [min′ər•əl] An object that is solid, is formed in nature, and has never been alive **(C6)**

mixture [miks′chər] A substance that contains two or more different types of matter **(E41)**

motion [mō′shən] A change in position **(F59)**

mountain [moun′tən] A landform; a place on Earth's surface that is much higher than the land around it **(C35)**

nonrenewable resource [non′ri•nōō′ə•bəl rē′sôrs] A resource, such as coal or oil, that will be used up someday **(C96)**

orbit [ôr′bit] The path an object takes as it moves around another object in space **(D58)**

GLOSSARY

Multimedia Science Glossary: www.harcourtschool.com/scienceglossary

phases [fāz•əz] The different shapes the moon seems to have in the sky when observed from Earth **(D76)**

photosynthesis [fōt′ō•sin′thə•sis] The food-making process of plants **(A20)**

physical change [fiz′i•kəl chānj] A change to matter in which no new kinds of matter are formed **(E40)**

physical property [fiz′i•kəl prop′ər•tē] Anything you can observe about an object by using your senses **(E6)**

plain [plān] A landform; a flat area on Earth's surface **(C35)**

planet [plan′it] A large body of rock or gas that orbits the sun **(D58)**

plateau [pla•tō′] A landform; a flat area higher than the land around it **(C35)**

population [pop′yoo•lā′shən] A group of the same kind of living thing that all live in one place at the same time **(B7)**

precipitation [prē•sip′ə•tā′shən] The water that falls to Earth as rain, snow, sleet, or hail **(D18)**

predator [pred′ə•tər] An animal that hunts another animal for food **(B54)**

prey [prā] An animal that is hunted by a predator **(B54)**

prism [priz′əm] A solid, transparent object that bends light into colors **(F44)**

producer [prə•doos′ər] A living thing that makes its own food **(B43)**

recycle [rē•sī′kəl] To reuse a resource to make something new **(C100)**

reflection [ri•flek′shən] The bouncing of light off an object **(F36)**

refraction [ri•frak′shən] The bending of light when it moves from one kind of matter to another **(F38)**

renewable resource [ri•noo′ə•bəl rē′sôrs] A resource that can be replaced in a human lifetime **(C94)**

reptile [rep′til] A land animal that has dry skin covered by scales **(A55)**

resource [rē′sôrs] A material that is found in nature and that is used by living things **(C88)**

revolution [rev′ə•loo′shən] The movement of one object around another object **(D68)**

rock [rok] A solid made of minerals **(C8)**

rock cycle [rok′ sī′kəl] The process in which one type of rock changes into another type of rock **(C14)**

root [root] The part of a plant that holds the plant in the ground and takes in water and minerals from the soil **(A7)**

R51

rotation [rō·tā′shən] The spinning of an object on its axis **(D68)**

S

salt water [sôlt′ wôt′ər] Water that has a lot of salt in it **(B26)**

scales [skālz] The small, thin, flat plates that help protect the bodies of fish and reptiles **(A52)**

sedimentary rock [sed′ə·men′tər·ē rok′] A rock formed from material that has settled into layers and been squeezed until it hardens into rock **(C12)**

seed [sēd] The first stage in the growth of many plants **(A12)**

seedling [sēd′ling] A young plant **(A13)**

simple machine [sim′pəl mə·shēn′] A tool that helps people do work **(F70)**

soil [soil] The loose material in which plants can grow in the upper layer of Earth **(C62)**

solar eclipse [sō′lər i·klips′] The hiding of the sun that occurs when the moon passes between the sun and Earth **(D80)**

solar system [sō′lər sis′təm] The sun and the objects that orbit around it **(D58)**

solid [sol′id] A form of matter that takes up a specific amount of space and has a definite shape **(E11)**

solution [sə·lōō′shən] A mixture in which the particles of two different kinds of matter mix together evenly **(E42)**

speed [spēd] The measure of how fast something moves over a certain distance **(F61)**

star [stär] A hot ball of glowing gases, like our sun **(D84)**

stem [stem] A plant part that connects the roots with the leaves of a plant and supports the plant above ground; it carries water from the roots to other parts of the plant **(A7)**

strip cropping [strip′ krop′ing] A type of planting that uses strips of thick grass or clover between strips of crops **(C76)**

telescope [tel′ə·skōp′] An instrument used to see faraway objects **(D88)**

temperature [tem′pər·ə·chər] The measure of how hot or cold something is **(D36)**

thermal energy [thûr′məl en′ər·jē] The energy that moves the particles in matter **(F7)**

thermometer [thûr·mom′ə·tər] A tool used to measure temperature **(F20)**

topsoil [top′soil′] The top layer of soil made up of the smallest grains and the most humus **(C63)**

trait [trāt] A body feature that an animal inherits; it can also be some things that an animal does **(A38)**

tropical rain forest [trop′i•kəl rān′fôr′ist] A hot, wet forest where the trees grow very tall and their leaves stay green all year **(B14)**

valley [val′ē] A landform; a lowland area between higher lands, such as mountains **(C35)**

volcano [vol•kā′nō] An opening in Earth's surface from which lava flows **(C49)**

volume [vol′yo͞om] The amount of space that matter takes up **(E22)**

water cycle [wôt′ər sī′kəl] The movement of water from Earth's surface into the air and back to the surface again **(D19)**

weather [weth′ər] The happenings in the atmosphere at a certain time **(D32)**

weather map [weth′ər map′] A map that shows weather data for a large area **(D46)**

weathering [weth′ər•ing] The process by which rock is worn down and broken apart **(C40)**

weight [wāt] The measure of the pull of gravity on an object **(F62)**

wind [wind] The movement of air **(D40)**

work [wûrk] The measure of force that it takes to move an object a certain distance **(F66)**

Abdominal muscles, R36
Absorption, F40
Activity pyramid, weekly plan, R22
Aerobic activities, R26–27
African elephant, A44
Agricultural chemist, A26
Agricultural extension agent, A25
Air
    animal need for, A35
    as gas, E12
    as resource, C89
Air mass, D36
Air pressure, D30
Akubo, Akira, B60
Alligators, A35, A56
Aluminum, C7
    recycling, C100, C101, C102–103
American egret, B30
Ammonites, C21
Amphibians, A50–51
Anasazi people, C10, D90
Anemometer, D40
Anemones, B27
Angler fish, B29
Animals
    adaptation of, for hunting, B44
    discovering, A58–59
    grouping, A58
    needs of, A34–38
    and people, B58–59
    sense of smell (chart), E3
    taming of, B58
    traits of, A38
    types of (chart), A31
    uses of, B58–59

Ants, C64
Anza Borrego Desert, CA, B20
Aristotle, A58–59, E30
Arteries, R40, R41
Asteroids, D64
Astronaut, D92
Astronomy, history of, D90
Atmosphere, D30
    layers of, D31
Atom(s), E16
Atomic bomb, E52
Atomic theory, E30
Axis, Earth's, D68, D71, D87

Bacon-Bercey, Joan, D50
Bacteria, B43, F9
Balance, using, R6
Balanced forces, F60
Bananas, A22
Barnacles, B27
Barn owl, A46
Barrel cactus, B21
Barrier island, C35–36
Basic light colors, F46
Bauxite, C7, C100, C103
Beaks, A46
Bears, A36
Beaufort wind scale, B40
Beaver dams, A37
Bedrock, C63
Beech tree leaf, A8
Bees, B4
Bell, Alexander Graham, F50
Biceps, R36, R37
Bicycle, safety, R12, R13
Big Dipper, D84, D86–87
Big-horned sheep, B43
Binoculars, D89
Biomedical engineer, D92

Biotite, C8
Bird(s)
    feet of, A46
    grouping, A46
    sizes of, A45
    traits of, A45
Bird banding, B34
Blackland Prairie, TX, B32
Black oak leaf, A8
Blizzards, D32
Body Systems, Human, R32–45
Brain, R44, R45
Breakage, E10
Breathing, R43
Breccia, C15
Bricks, programmable, F74–75
Buckland, William, C24
Bulbs, A14
Bullhead catfish, B30
Burning, E47, F9
Butterfly fish, A30

Cactus
    largest, A3
    parts of, B21
Camels, B59
Canopy, B15
Canyon, C36
Capillaries, R41
Carbon dioxide, A20
Cardinal, A46
Cardinal fish, A54
Cartilage, R35
Carver, George Washington, A26
Cassiopeia, D85
Casts, fossil, C21

INDEX

Cat and kittens, A42–43
Cave paintings, B58
Celery, A22
Celsius scale, F21
Chalk, C6
Chameleon, A36
Cheetah cubs, A38
Chemical changes, using, E48
Chesapeake Bay, D12
Chinook, D26
Chlorophyll, A20
Circulatory system, R40–41
    caring for, R41
Civil engineer, E51
Clavicle (collar bone), R34
Clay, C69
Clouds, D17
Coal, C96
Coastal forests, B15
Cocklebur, A16
Cold front, D46
Color, E7, F44–47
    adding, F46
    of minerals, C6
Colorado River, C36
Comets, D64, D65
Communities, B7
Compact discs (CDs), F49
Composite plastic, E51
Compost, C60, F9
Compost Critters (Lavies), C65
Computer, using, R8–9
Concave lenses, F48
Condensation, D17, E41
Conductors, F15
Cone, evergreen, A12, A14
Coniferous forests, B16
Conservation, soil, C74–77
    defined, C76
Constellations, D85
Construction worker, C53

Consumers, B43, B49
Continent(s), D10–11
Continental glaciers, C44
Contour plowing, C72, C76
Cooter turtle, B30
Copernicus, D90
Copper, C7, C91
Copper sulfate, E42
Core, Earth's, C8
Corn seedling, A10
Corundum, C6
Creep, defined, C42
Crescent moon, D77
Cricket chirp thermometer, F3
Crocodiles, A56
Crop rotation, A26
Crust, Earth's, C8
Crystals, C9
Cumulus clouds, D38
Cuttings, A14

D

Dalton, John, E30
Darden, Christine, F76
Day and night, causes of, D72
Deciduous forests, B13
Decomposers, B44
Deer, A40
Delta, C43
Deltoid, R36
Democritus, E30
Desert(s)
    animals of, B22
    hot ground of, F16
    plants of, B21
    types of, B20
Devil's Tower, C32
Dew, D17

Dial thermometer, F20
Diamonds, C6, C7
Diaphragm, R42, R43
Digestive system, R38–39
    caring for, R38
Dinosauria, C24
Dinosaurs, discovering, C24–25
Dirzo, Rodolfo, A60
Distance, F61
Dogs and puppies, A43
Dolphins, B24
Doppler radar, D45
Drought-resistant plants, A24–25
Duck and ducklings, A39
Duckweed, A2
Dunes, C30, C43

E

Ear(s), caring for, R32
Ear canal, R32
Eardrum, R32
Earth, D58–59
    facts about, D61
    layers of, C8
    movement of, D68–69
    seasons of, D71
    sunlight on, D72
    surface of, C34–37, D70
    tilt on axis of, D70
    water of, D6–13
Earthquake(s), C48
    damage from, C32
    in Midwest (chart), C31
Earthquake-proof buildings, C52–53

R55

Earthquake safety, R15
Earthworms, C64
Eastern box turtle, A55
Echidna, spiny, A44
Ecologist, C80
EcoSpun™, C107
Ecosystem(s)
    changes in, B8–9
    dangers to, B8–9
    desert, B20–23
    estuaries of, D12
    field guide, B33
    forest, B12–17
    home movies of, B22–33
    marsh, B54–55
    parts of, B7
    pond's, B7
    water, B26–31
Edison, Thomas, F50
Edison Pioneers, F50
Eggs
    bird, A45
    fish, A54
    reptile, A55
Element, E12, E30
Elodea, A18
Emerald, C6
Endangered animals (chart), B3
Endangered Species Act, B58
Endeavour **mission,** D92
Energy
    conservation of, C101
    defined, F6
    from food, B50
    from sunlight, B49
    transfer of, B42, B48, B50
Energy pyramid, B50
Engineer, F50, F76

Environment(s)
    computer description of, B32–33
    and living things, B6
Environmental technician, D21
Equator, D11
Erosion, C42–43
Esophagus, R38, R39
Estuary(ies), D12
Evaporation, D17, D18, E18, E41
Exhaling, R43
Extremely Weird Hunters **(Lovett),** B57
Eye(s), caring for, R32

F

Fahrenheit scale, F21
Farmer, C79
Farming
    with computers, C79
    with GPS, C78–79
Feathers, A45
Feldspar, C8
Femur, R34
Fermi, Enrico, E52
Fertilizers, C70
Fibula, R34
Film, photographic, chemical changes to, E48
Fins, A53
Fire, and thermal energy, F8
Fire safety, R14

First aid
    for bleeding, R17
    for burns, R18
    for choking, R16
    for insect bites, R19
    for nosebleeds, R18
    for skin rashes from plants, R19
First quarter moon, D77
Fish, A52–54
Fish schools, B60
Flake, plastic, C106
Flames, energy in, F6
Flexors, R36
Floods, C50, D38
Floor, forest, B15
Flowers, A14
Food
    animal need for, A36
    and bacteria, R30
    and energy, B48–51
    groups, R28
    plants use of, A20–22
    pyramid, R28
    safety tips, R31
Food chains, B48–49
Food webs, B54–57
Force(s), F58
Forest(s), types of, B12–17
Forest fires, B8
Forklift, F64
Fossils
    defined, C18, C20
    dinosaur, C24, F25
    dragonfly, C21
    fish, C3
    formation of, C20–21
    plant, C21
    types of, C20
Franklin, Benjamin, F48
Fresh water, D8–9

INDEX

Freshwater ecosystem, B26, B30
Frilled lizard, Australian, A57
Frog, metamorphosis of, A51
Fulgurite, C2
Full moon, D77
Fungi, B43

Galápagos tortoise, A56
Galilei, Galileo, D88, D90–91, F48
Gas, E11–12
   particles of, E17
Gasoline, C90
Gel, E10
Geochemist, C26
Geode, C4
Geologist, C54, D22
Germination, A13
Giant pandas, B43
Gibbous moon, D77
Gill(s), A51, A52, A53
Gilliland Elementary School, Blue Mound, TX, B22–33
Glaciers, C44–45, D8
   longest, C45
Global Learning and Observations to Benefit the Environment (GLOBE), B33
Global Positioning System (GPS), C78–79
Gneiss, C13
Gold, C6, C7
Goldfish, A53
Gorillas, A43
Graduate, E26
Granite, C9, C13

Graphite, C6, C7
Gravity, Earth's, F62
   of sun, D59
Great blue heron, A46
Great Red Spot, D62
Great Salt Lake, UT, D10
Green boa, A56
Ground cone, B2
Groundwater, D8, D18
Grouping mammals, A44
Guadeloupe bass, A52

Habitats, B7
Hail, D39
Halley's comet, D65
Hand lens, using, R4
Hardness, of minerals, C4, C6
Hawkins, Waterhouse, C25
Heart, R40, R41
   aerobic activities for, R26–27
Heat
   defined, F8
   and matter, E18
   and physical change, E38, E42
   and water forms, D16–17
High pressure, D46
Hippopotamus, B39
Hodgkin, Dorothy Crowfoot, E32
Home alone safety measures, R20–21
Hooke, Robert, A59
Hot-air balloons, D30, F2
"Hot bag," F24–25
Hubble Space Telescope, D91
Humerus, R34

Humus, C62, C68
Hurricanes, B41, D32
   record (chart), D27
Hyena, B51

Ice, D16, E11
Iceberg, D8
Icecaps, D8
Igneous rock, C12–13, C17
Iguanodon, C24
Imprints, fossil, C21
Inclined plane, F70–71
Indian paintbrush, A6
Inexhaustible resources, C94–95
Inhaling, R43
Inheritance, traits, A38
Inner ear, R32
Inner planets, D60–61
Inside the Earth (Cole), C9
Insulators, F15
Insulin, E32
Interaction, of plants and animals, B42
Internet, R9
Inventor, F26, F50
Iris, eye, R32
Iron, C7
   as conductor, F15
Irrigation, D7

Jackrabbit, B23
Jade plant leaf, A8

R57

Jemison, Mae, D92
Jefferson Memorial, C16
Jupiter, D58–59
    facts about, D62

Keck telescope, D91
Kelp, B28
Kitchen, cleanliness, R31
Kits, beaver, A37
Koala, A44

Ladybug, A32
Lake(s), B30
Lake Baikal, Russia, C84
Lake Michigan, C43
Lake Saint Francis, AR, C32
Lakota people, D85
Landfills, C100, C101
Landforms, C34–37
    defined, C34
Landslides, Slumps, and Creep (Goodwin), C45
Langmuir, Charles, C26
Large intestine, R38
Lasers, F49
    and lightning, D39
Latimer, Lewis Howard, F50
Lava, C46, E18
Lava temperatures, C3
Layers, in rain forest, B14–15
Leaves, A7
    shapes of, A8
Lemon tree, A18
Lens, eye, R32
Lenses, F48
Lever, F70–71

Light
    bending, F38
    bouncing, F36–37
    and color, F44–45
    direction of, F38–39
    energy, F34
    speed of, F30, F31, F38, F44
    stopping, F40
Light and optics discovery, F48
Lightning, D48–49
Limestone, C20, C68
Lippershey, Hans, D90
Liquid(s), E11–12
    measuring, R7
    particles of, E17
Live births, A42–43
    fish, A52
    reptile, A55
Liver, R38
Living things, and food, B42–45
Lizards, A56
    described, C70
Loam, C68
Lodge, beaver, A37
Look to the North: A Wolf Pup Diary (George), A39
Low pressure, D46
Lunar eclipse, D78–79
Lungs, A35, A42, A45, A51, A57, R42, R43
    aerobic activities for, R26–27

Machine(s)
    compound, F72
    simple, F70–73
Magma, C8, C13

Magnet, E10
Maiman, T.H., F48
Mammals, A42–44
    types of, A44
Mangrove trees, D13
Mangrove Wilderness (Lavies), B51
Mantell, Gideon and Mary Ann, C24
Mantle, Earth's, C8
Mapmaking, on Internet, B33
Marble, C14, C16, C92
Mars, D58–59, F58, F62
    erosion of, D22
    facts about, D61
Mass
    adding, E27
    defined, E24
    measuring, E24–25
    of selected objects, E25
    and volume compared, E28
Matter
    appearance of, E7
    changing states of, E18
    chemical changes to, E46–49
    defined, E6
    history of classifying, E30
    measuring, E22–26
    physical changes in, E39–43
    physical properties of, E6–13
    states of, E11–12
Mayfly, B30
Measuring cup, E22
Measuring pot, E26
Measuring spoon, E26
Medeleev, Dmitri, E30
Megalosaurus, C24

INDEX

Mercury, D58–59
   facts about, D60
Mesosphere, D31
Metamorphic rock, C12–13, C16, C17
Meteor(s), D64
Meteorites, D64
Meteorologists, D32, D38, D46
Meterstick, using, R7
Microscope, A58–59, E16
   invention of, F48
   using and caring for, R5
Microwave ovens, F26
Middle ear, R32
Mineral(s)
   defined, C6
   in salt water (chart), D10
   in soil, C62, C63
   use of, C7
Mining, C90
Mirrors, F37
Mission specialist, D92
Mississippi Delta, C42
Mississippi River, C43
Mixtures, E41
   kinds of (chart), E36
Models, scientific use of, C19
Molds, fossil, C21
Montserrat Island, C49
Moon, Earth's
   and Earth interaction, D76–81
   eclipses of, D78–79
   phases of, D76–77
   rotation and revolution of, D69, D76
   weight on, F62
Moon(s), planetary, D62
Moon "rise," D74
Moose, B10
Moths, B14
Motion, F59–60
Mountain, C35
Mount Everest, Himalayas, E2
Mount Palomar observatory, D90
Mouth, R38, R39, R42, R43
Mud flows, C49
Mudslides, C50
Muscle pairs, R37
Muscovite, C8
Muscular system, R36–37
   caring for, R36
Mushroom Rock, C40

N

NASA, D21, D22, F76
Nasal cavity, R33
National Aeronautics and Space Administration. See NASA
National Oceanic and Atmospheric Administration, D50
National Weather Service, D45, D50
Natural gas, C97
Negative, photographic, E48
Neptune, D58–59
   facts about, D63
Nerves, R44
Nervous system, R44–45
   caring for, R45
New Madrid Fault, C32
New moon, D77
Newt(s), A50
Newton, Isaac, F48
Nice, Margaret Morse, B34
Nobel Prize in Chemistry, E32
Nobel Prize in Physics, E52
Nonliving things, in environment, B7
Nonrenewable resources, C96–97
Northern spotted owl, B15
North Star, D87
Nose, R42, R43
   caring for, R33
Nostril, R33

O

Obsidian, C15
Ocean, resources of, D11
Ocean ecosystems, B28–29
Ocean food web, B56
Oil
   as resource, C90
   search for, C108
Oil derrick, C90
Oil pump, C86
Oil wells, D11
Old-growth forests, C96
Olfactory bulb, R33
Olfactory tract, R33
Olympus Mons, D61
Opacity, F40
Optical fiber telephone, F48–49
Optic nerve, R32
Optics, history of, F48
Orangutan, A44
Orbit, D58
Ordering, scientific, C5
Organization for Tropical Studies, A60
Orion, D85

Ornithologist, B34
Outer Banks, NC, C36
Outer ear, R32
Outer planets, D60, D62–63
Owen, Richard, C24–25
Oxygen, A35, A52
    and plants, A21
    as resource, C88, C89

Pacific Ocean, D10
Pan balance, E24
Panda, A36, B53
Paper, changes to, E39
Partial lunar eclipse, D79
Partial solar eclipse, D80
Particles
    bumping, F14
    connected, E17
    in fire, F7
    and heat, F7
    in liquids and gases, F16
    of solids, F7
Patents, F50
Peace Corps, D92
Peanuts, A26
Pecans, A26
Pelvis, R34
People and animals, history of, B58
Periodic table, E31
Perseus, D85
Petrologist, C26
Photosynthesis, A20–21
Physical changes, kinds of, E39
Physical property, defined, E6
Physician, D92
Physicist, E52

Pill bugs, C64
Pine cone, largest, A3
Pizza chef, F25
Pizza holder, as insulator, F15
Plain, C35
Planets, D58
    distance from sun, D58–59
    facts about, D57
Plankton, B60
Plant(s)
    drought-resistant, A24–25
    foodmaking, A20–21
    heights of (chart), A2
    need for light, F34
    needs of, A6–9
    parts of, A7
    seed forming, A14
    use of food by, A22–23
    and weathering, C41
Plant material, in different environments (chart), B38
Plastic, C90, C100
    bridges from, E50–51
    recycling of, C106–107
Plateau, C35
Pluto, D58–59
    facts about, D63
Polar bear, A42
Polaris, D87
Pollution, D8, D40, C101
    preventing, C104
Polo, Marco, F48
Ponds, B30
Populations, B7
Postage scale, E26
Potatoes, A22
Potting soil, C70
Precipitation, D8, D17, D18, D32
    measuring, D38
Predator, B54

Prey, B54
Prisms, F44
Producers, B43, B49
Property, defined, C4
Puffin, B46
Pull, F58
Pulley, F70–71
Pumice, C15
Pupil, eye, R32
Purple gallinule, A46
Push, F58, F60

Quadriceps, R36
Quartz, C6, C7, C16
Quartzite, C14
Quinones, Marisa, C108

Raccoon, B6
Radiation, F16
Radio telescope, Arecebo, D91
Radishes, A4
Radius, R34
Rain, D32
Rainbow(s), F44
    formation of, F45
Rainbow trout, A53
Rainfall, by type of forest, B12
Rain forests, C75
    See also Tropical rain forests
Rain gauges, D38
Ray, John, A58–59

INDEX

Recycling, C100–103
    plastics, C106–107
Recycling plant worker, C107
Red sandstone, C15
Reflection, F36
Refraction, F38, F44
Renewable resources, C94
Reptiles, A55–57
Resources
    conserving, C104
    daily use of Earth's, C85
    defined, C74
    kinds of, C94–97
    location of, C88–89
    under Earth's surface, C90
Respiratory system, R42–43
    caring for, R43
Retina, R32
Revolution, Earth's, D68
Rhinoceros, B3
Rib, R35
Rib cage, R34, R35
Rivers, B30, C35, C50
Rock(s)
    described, C8
    formation of, C12–15
    types of, C12–15
    use of, C16–17
    weathering of, C40–41
Rock, The (Parnall), C17
Rock cycle, C14–15
Rocky Mountains, C32
Roots, A7, A14
Rose, fragrance of, E9
Rossbacher, Lisa, D22
Rotation, Earth's, D68
Rowland, Scott, C54
Rubbing, and thermal energy, F8
Ruler, using, R7
Rusting, E47

S

Safety
    and health, R12–21
    in Science, xvi
Saguaro cactus, B22
Salamander, A50
Salt water, D10–11
Saltwater ecosystems, B26, B27–28
San Andreas Fault, CA, C48
Sand, C69
Sandstone, C15
Sapphire, C6
Satellite, D45
Saturn, D58–59
    facts about, D62
Scales, fish, A52, A53
Scavenger, B51
Schist, C13
Scissors, operation of, F72
Scorpion, B18
Screw, F70–71
Sea horse, A38
Sea otter, B52
Seasons, D68–73
    causes of, D70
Sediment, C12
Sedimentary rock, C12–13
Seed(s)
    berry, A16
    described, A12
    kinds of, A14
    mangrove, A16
    milkweed, A17
    needs of, A13
    parts of, A15
    sizes of, A15
    spreading, A16–17
    sunflower, A12
Seed coat, A15
Seedling, A13, A15
Segre, Emilio, E30
Sense(s), describing matter through, E7–10
Sense organ(s), R32–33
Serving sizes, R29
Shadows, D69, F32, F35
Shale, C20
Shape, of minerals, C6
Sharks, A52
    sense of smell of, E37
Shelter, animal need for, A37
Sidewinder, B22
Silicon chips, C16
Size, E7
Skeletal system, R34–35
    caring for, R35
Skin, R33
    layers of, R33
Skull, R34, R35
Skunk, odor of, E9
Sky watchers, D90–91
Slate, C13, C14
Small intestine, R38
Snakes, A38, A56, B22
Snow boards, D38
Snowflake porphyry, C15
Soil
    conservation of, C74–77
    formation of, C62–63
    importance of, C64–65
    life in, C64
    living things in (chart), C59
    of Mars, C59
    minerals in, A7
    parts of, C70
    pigs use of, C58
    as resource, C89
    types of, C68–71

R61

Soil roundworms, C80
Solar eclipse, D80
Solar energy, F35
Solar system, D58–65
   structure of, D58
Solid(s), E11
   particles of, E17
Solution, E42
   to solid, E46
Sonic boom, F76
Speed, F54, F60–61
   of different objects (chart), F55
Spencer, Percy, F26
Spiders, B8
Spinal cord, R44, R45
Spine, R34
Spring scale, E26, F56, F58
   using, R6
Squash, growth of, A13
Standard masses, R6
Star(s)
   defined, D84
   and Earth's movement, D86–87
   observing, D84–89
   patterns, D84–85
   sizes of (chart), D55
Stars: Light in the Night, The (Bendick), D89
Stem, A7
Sternum, R35
Stickleback, A54
Stomach, R38, R39
Stonehenge, D90
Stored food, seed, A15
Storm clouds, D28
Storm safety, R15
Stratosphere, D31
Stratus clouds, D38
Strawberries, A22
Streams, B30

Stretches, R24–25
Strip cropping, C26
Strip mining, C90, C103
Subsoil, C63
Sugar, plant, A21
Sun
   eclipse of, D80
   heat from, F16
   light energy of, A20–21
   as star, D59
Sun and Moon, The (Moore), D81
Sweet gum leaf, A8
Sweet potatoes, A26

T

Tadpoles, A49, A51
Taklimakan desert, China, B20
Taste buds, R33
Tayamni, D85
Teacher, B33
Technetium, E30
Telescopes, D56, D82, D88, D91, D95
Temperature, D32, E8
   measuring, D36, F20–23
   by type of forest, B12
   various measurements of, F20–21
Tendon(s), R37
Texas bluebonnets, A6
Thermal energy, D16, F6–11, F24
   controlling, F22
   movement of, F14–17
   producing, F9
   use of, F10
Thermal retention, F24–25
Thermal vents, A59

Thermometer(s), F18, F21
   using, R4
   working of, F20
Thermosphere, D31
Thermostat, F23
   working of, F22
Third quarter moon, D77
Thunderhead, D28
   composition of, D39
Tibia, R34
Tilling, C76
Timing device, using, R7
Toads, A50
Tongue, R33
Tools, of measurement, E26
Topsoil, C63, C70
Tornadoes, B41, D33, D39
Tortoises, A56
Total lunar eclipse, D79
Total solar eclipse, D80
Toy designer, F75
Trachea, R42
Traits, animal, A38, A44
Translucency, F40
Transparency, F40
Trash, C101
Tree frog, A48
Triceps, R36, R37
Triton, temperatures on, D55
Tropical ecologist, A60
Tropical fish, B29
Tropical rain forests, B14–15
   See also Rain forests
Troposphere, D31
Tubers, A14
Turtles, A56, B48
Tuskegee Institute, AL, A26
Tyrannosaurus rex, C18

Ulna, R34
Umpqua Research Company, D21
Understory, B15
Uranus, D58–59
   facts about, D63
Ursa Major, D84

Valles Marineris, Mars, D54
Valley, C35
Valley glaciers, C44
Valve, R41
Veins, R40, R41
Venus, D58–59
   facts about, D60
Verdi **(Cannon),** A57
Very Large Array telescope, D95
Volcanoes, C13, C46, C49–50, E18
   Hawaiian, C54
   types of, C54
Volume
   defined, E22
   and mass compared, E28
   measuring, E22–23

Wall, Diana, C80
Warm front, D46
Waste (chart), C100

Water
   amounts of salt and fresh, D7
   forms of, D16–17
   importance of, D6–7
   as resource, C88
   in space, D21
   uses of, D3
   wasting, D2
   and weathering, C41
Water cycle, D16–19
Water filters, D20–21
Water hole, African, A35
Water treatment plant, D11
Water vapor, D16, D17
Weather, D32
   gathering data, D44–45
   measuring, D36–41
Weather balloons, D45
Weather forecasting, D44–47
Weather fronts, D37, D46
Weather maps, D46
Weather researcher, D49
Weather satellites, D44
Weather station, D44
   symbols used by, D46, D47
Weather vanes, D34
Weathering, C40–41
Weaverbird, A45
Wedge, F70, F72
Weighing scale, E20
Weight, F62
Wells, C90
West Chop Lighthouse, MA, C31
Wetness, E8
Whales, A44
Wheel and axle, F70, F72
White light, F46
   colors of, F45
Wildfires **(Armbruster),** B9

Wind, D32
   measuring, D40
Wings, A45
Women in Science and Engineering (WISE) Award, F76
Work, F66–67
Workout, guidelines for, R23–25
World Health Organization, D20–21
Wright, James, E37

Xeriscaping, A25

Yellowstone National Park, B8
Yellow tang, A53

Zinc, C7

Photography Credits - Page placement key: (t) top, (c) center, (b) bottom, (l) left, (r) right, (bg) background, (i) inset

Cover Background, Charles Krebs/Tony Stone Images; Inset, Jody Dole.

Table of Contents - iv (bg) Thomas Brase/Tony Stone Images; (i) Denis Valentine/The Stock Market; v (bg) Derek Redfearn/The Image Bank; (i) George E. Stewart/Dembinsky Photo Association; vi (bg) Richard Price/FPG International; (i) Martin Land/Science Photo Library/Photo Researchers; vii (bg) Pal Hermansen/Tony Stone Images; (i) Earth Imaging/Tony Stone Images; viii (bg) Steve Barnett/Liaison International; (i) StockFood America/Lieberman; ix (bg) Simon Fraser/Science Photo Library/Photo Researchers; (i) Nance Trueworthy/Liaison International.

Unit A - A1 (bg) Thomas Brase/Tony Stone Images; (i) Denis Valentine/The Stock Market; A2-A3 Joe McDonald/Bruce Coleman; A3 (i) Marilyn Kazmers/Deminsky Photo Associates; A4 Ed Young/AGStock USA; A6 (l) Anthony Edgeworth/The Stock Market; (r) Chris Vincent/The Stock Market; A6-A7 (bg) Barbara Gerlach/Dembinsky Photo Associates; A7 (c) Wendy W. Cortesi; A8 (t) Runk/Schoenberger/Grant Heilman Photography; (c) Runk/Schoenberger/Grant Heilman Photography; (bl) Renee Lynn/Photo Researchers; (br) Dr. E.R. Degginger/Color-Pic; A9 Runk/Schoenberger/Grant Heilman Photography; A10 Runk/Schoenberger/Grant Heilman Photography; A12 (l) Bonnie Sue/Grant Heilman/Photo Researchers; (li) Klaus Paysan/Peter Arnold, Inc.; (r) Runk/Schoenberger/Grant Heilman Photography; A13 (t) Ed Young/AgStock USA; (b) Dr. E. R. Degginger/Color-Pic; A14 (t) Richard Shiell/Dembinsky Photo Associates; (b) Robert Carr/Bruce Coleman, Inc.; (br) Scott Sinklier/AGStock USA; A16 (t) Thomas D. Mangelsen/Peter Arnold, Inc.; (c) E.R. Degginger/Natural Selection Stock Photography; (b) Randall B. Henne/Dembinsky Photo Associates; (l) Stan Osolinski/Dembinsky Photo Associates; (l) Scott Camazine/Photo Researchers; A17 William Harlow/Photo Researchers; A18 Christi Carter/Grant Heilman Photography; A20 Runk/Schoenberger/Grant Heilman Photography; A22 (l) DiMaggio/Kalish/The Stock Market; (b) Jan-Peter Lahall/Peter Arnold, Inc.; (br) Holt Studios/Nigel Cattlin/Photo Researchers; A23 Robert Carr/Grant Heilman Photography; A24 Richard Shiell; A25 J. Sapinsky/The Stock Market; A26 (b) Corbis; A30-A31 (bkgd) Art Wolfe/Tony Stone Images; A31 (cr) Astrid & Hanns Frieder Michler/Science Photo Library/Photo Researchers; A32 (t) Rosemary Calvert/Tony Stone Images; (1) Ralph A. Reinhold/Animals Animals; (2) Johnny Johnson/ Tony Stone Images; (3) Mike Severns/ Tom Stack & Associates; (4) Fred Whitehead/Animals Animals; (5) Art Wolfe/ Tony Stone Images; (6) J.C. Stevenson/Animals Animals; A34-A35 Doug Perrine/Innerspace Visions; A35 (t) Ronald Hellstrom/Bruce Coleman, Inc.; (b) Stan Osolinski/Tony Stone Images; (br) Mike Severns/Tony Stone Images; (lc) Kevin Schafer/Tony Stone Images; (bl) Marilyn Kazmers/Peter Arnold, Inc.; Keren Su/Tony Stone Images; A38 (t) Rudie Kuiter/Innerspace Visions; (c) Fred Bruemmer/Peter Arnold, Inc.; (b) Art Wolfe/Tony Stone Images; A39 Phil A. Dotson/Photo Researchers; A40 Brian Stablyk/Tony Stone Images; A43 (b) Paul Metzger/Photo Researchers; (b) Frans Lanting/Minden Pictures; A44 (t) Stephen Dalton/Photo Researchers; (c) Tom McHugh/Photo Researchers; (c) Evelyn Gallardo/Peter Arnold, Inc.; (bl) The Photo Library-Sydney/Gary Lewis/Photo Researchers; (br) Francois Gohier/Photo Researchers; A45 (l) Theo Allofs/Tony Stone Images; (blue jay) Wayne Lankinen/Bruce Coleman, Inc.; (macaw) Ms. Mastrorillo/The Stock Market; (emperor penguin) Kjell B. Sandved/Photo Researchers; (ostrich) Leonard Lee Rue III/Photo Researchers; (bee humming bird) Robert A. Tyrrell Photography; (peacock) Tom McHugh/Photo Researchers; A46 (t) Manfred Danegger/Tony Stone Images; (cl) John Cancalosi/Peter Arnold, Inc.; (b) Bill Ivy/Tony Stone Images; (br) Stan Osolinski/The Stock Market; A48 (t) O.S.F./ Animals Animals; (b) Tim Davis/Tony Stone Images; A50 (tl) Nuridsany et Perennou/Photo Researchers; (c) E.R. Degginger/Color-Pic; A52 (t) Thomas D. Collins/Photo Researchers; A52 (t) David M. Schleser/Nature's Images; (c) Andrea & Antonella Ferrari/Innerspace Visions; A52-A53 Kelvin Aitken/Peter Arnold, Inc.; A53 (t) Zig Leszczynski/Animals Animals; (c) Kelvin Aitken/Peter Arnold, Inc.; (r) Tom McHugh/Steinhart Aquarium/Photo Researchers; A54(l) Kim Taylor/Bruce Coleman, Inc.; (c) Fred Bavendam/Minden Pictures; A55 (l) Fred Bavendam/Minden Pictures; A55 (r) Zig Leszczynski/Animals Animals; (c) Suzanne L. Collins & Joseph T. Collins/Photo Researchers; (bli) Dwight R. Kuhn; A56 (t) Jany Sauvanet/Photo Researchers; (c) G.E. Schmida/Fritz/Bruce Coleman, Inc.; A56-A57 (b) Tom & Pat Leeson/Bruce Coleman, Inc.; A57 Schafer & Hill/Tony Stone Images; A58 (t) Tom Brakefield/Bruce Coleman, Inc.; (c) Dr. E.R. Degginger/Color-Pic; (b) Michael Holford; A59 Emory Kristof/National Geographic Image Collection; A60 (tr) Bertha G. Gomez; (bl) Michael Fogden/bruce Coleman, Inc.

Unit B - B1 (bg) Derek Redfearn/The Image Bank; (i) George E. Stewart/Dembinsky Photo Association; B2-B3 (bg) Sven Linoblad/Photo Researchers; B2 (i) Wayne P. Armstrong; B4 Hans Pfletschinger/Peter Arnold, Inc.; B6 (t) Dwight R. Kuhn; (r) Michael Durham/ENP Images; B7 Frank Krahmer/Peter Arnold, Inc.; B8 (t) Jeff and Alexa Henry/Peter Arnold, Inc.; (b) Jeff and Alexa Henry/Peter Arnold, Inc.; (c) Christoph Burki/Tony Stone Images; B10 Kennan Ward/The Stock Market; B13 (all) James P. Jackson/Photo Researchers; B14 Zefa Germany/The Stock Market; B15 Janis Burger/Bruce Coleman, Inc.; B16 (r) Michael Quinton/Minden Pictures; B16-B17 (b)Grant Heilman/Grant Heilman Photography; B18 J.C. Carton/Bruce Coleman, Inc.; B20 (l) Wolfgang Kaehler Photography; (r) James Randklev/Tony Stone Images; B21 Dr. E.R. Degginger/Color-Pic; B22 (l) Paul Chesley/Tony Stone Images; (c) Jeff Foott/Bruce Coleman, Inc.; (i) Jen & Des Bartlett/Bruce Coleman, Inc.; B23 Lee Rentz/Bruce Coleman, Inc.; B24 Leo De Wys Inc.; B27 (li) R.N. Mariscal/Bruce Coleman, Inc.; (b) Dr. E.R. Degginger/Color-Pic; (i) Naitar E. Harvey, APSA/National Audubon Society/Photo Researchers; B28 Flip Nicklin/Minden Pictures; B29 (t) Norbert Wu/Peter Arnold, Inc.; (b) Norbert Wu/Peter Arnold, Inc.; B30 (t) Gary Meszaros/Peter Arnold, Inc.; (bli) Stevan Stefanovic/Okapia/Peter Arnold, Inc.; (bci) Dwight R. Kuhn; (bri) Phil Degginger/Color-Pic; (bl) Jeff Greenberg/Photo Researchers; B32 (b) Courtesy of Jane Weaver/Parie Project/L. A. Gilillard Elementary; (ti) Globe-NASA/ Goddard Scientific Visualization Studio; B33 Derke/O'Hara/Tony Stone Images; B34 (tr) The Marjorie N. Boyer Trust; (b) Anthony Merciexa/ Photo Researchers; B38-B39 Luiz E. Marigo/Peter Arnold, Inc.; B39 (r) Roland Seitre/Peter Arnold, Inc.; B40 (l) Roy Morsch/The Stock Market; (r) Norbert Wu/Tony Stone Images; (c) Rosemary Calvert/Tony Stone Images; B41 (bl) Stan Osolinski/The Stock Market; (c) R. Kopfle/KOPFL/Bruce Coleman; (br) Michael Durham/ENP Images; B42 (bg) Hans Reinhard/Bruce Coleman, Inc.; (li) Dwight R. Kuhn; (ri) Dr. Paul A. Zahl/Photo Researchers; B43 (t) Wolfgang Kaehler Photography; (b) Rob Hadlow/Bruce Coleman, Inc.; B44 (t) Stephen Dalton/Photo Researchers; (c) Andrew Syred/Science Photo Library/Photo Researchers; (b) Stephen Krasemann/Tony Stone Images; B46 Laurie Campbell/Tony Stone Images; B48 Dwight R. Kuhn; B49 (t) Paul E. Taylor/Photo Researchers; (c) Holt Studios/Photo Researchers; (b) Breck P. Kent/Animals Animals; B51 Mitsuaki Iwago/Minden Pictures; B52 Erwin and Peggy Bauer/Bruce Coleman, Inc.; B54-B55 Michael Durham/ENP Images; B57 Jane Burton/Bruce Coleman, Inc.; B58 (cl) LASCAUX Caves II, France/Explorer, Paris/Superstock; (c) Fred Bruemmer/Peter Arnold, Inc.; (b) Tom Brakefield/Bruce Coleman, Inc.; B60 (tl) Leah Edelstein-Keshet/University of British Columbia; (bl) Fred McConnaughey/Photo Researchers.

Unit C Other - C1(bg) Richard Price/FPG International; (i) Martin Land/Science Photo Library/Photo Researchers; C2-C3 (bg) E. R. Degginger; C2 (bc) A. J. Copley/Visuals Unlimited; C3 (ri) Paul Chesley/Tony Stone Images; C4 (t) The Natural History Museum, London; C6 (tl), (ct), (cb) Dr. E. R. Degginger/Color-Pic; (b) E. R. Degginger/Bruce Coleman, Inc.; (bl) Mark A. Schneider/Dembinsky Photo Associates; C6-C7 (b) Chromosohm/Joe Sohm/Photo Researchers; C7 (b) Blair Seitz/Photo Researchers; (tri), (bli) Dr. E. R. Degginger/Color-Pic; C8 (t) Barry Runk/Grant Heilman/Photo Researchers; (ct), (cb) Dr. E. R. Degginger/Color-Pic; (b) Dr. E. R. Degginger/Color-Pic; (cb) Barry L. Runk/Grant Heilman/Photo Researchers; C8-9 (b) Robert Pettit/Dembinsky Photo Researchers; C10 Tom Bean/Tom & Susan Bean, Inc.; C12 Jim Steinberg/Photo Researchers; C12-C13 (bg) G. Brad Lewis/Photo Resource Hawaii; C14 (l), (r) Dr. E. R. Degginger/Color-Pic; C14 (br) Aaron Haupt/Photo Researchers; C15 (tl) Robert Pettit/Dembinsky Photo Associates; (tr) Charles R. Belinky/Photo Researchers; (bl), (bc), (br) Dr. E. R. Degginger/Color-Pic; C16 (t) Roger Du Buisson/The Stock Market; (c) Jay Mallin Photos; C16-C17 (b) Ed Wheeler/The Stock Market; C18 Stephen Wilkes/The Image Bank; C21 (t) William E. Ferguson; (bl) Kerry T. Givens/Bruce Coleman, Inc.; C22 (t) & (bl) AP Photo/Dennis Cook; (b) M. Timothy O'Keefe/Bruce Coleman, Inc.; C24 (t) Francois Gohier/Photo Researchers; (b) The Natural History Museum, London; C25 Stan Osolinski; C26 (tr) Jean Miele/Lamont-Doherty Earth Observatory of Columbia University; C30-C31 (b) John Warden/Tony Stone Images; C31 (b) Harold Nadeau/The Stock Market; C32 G. Alan Nelson/Dembinsky Photo Associates; C33 (t) Superstock; C34-35 (b) Darrell Gulin/Dembinsky Photo Associates; C35 (t) Michael Hubrich/Dembinsky Photo Associates; (b) Mark E. Gibson; C36 (t) Breck P. Kent/Earth Scenes; C36-37 (b) Paraskevas Photography; C38 Mark E. Gibson; C40 (bl) Dr. E. R. Degginger/Color-Pic; C40 (b) Mark A. Schneider/Dembinsky Photo Associates; (br-b) Rod Planck/Dembinsky Photo Associates; C41 (b) Michael Hubrich/Dembinsky Photo Associates; (c) John Gerlach/Dembinsky Photo Associates; C42 (t) Georg Gerster/Photo Researchers; (c) NASA Photo/Grant Heilman Photography; C42-C43 (b) C.C. Lockwood/Earth Scenes; C43 (t) Mark E. Gibson; C46 Ken Sakamoto/Black Star; C48 (l) David Parker/SPL/Photo Researchers; (l) AP/Wide World Photos; C49 (l) AP/Wide World Photos; (b) AP Photo/Wide World Photos; (bl) Will & Deni McIntyre/Photo Researchers; C50-C51 AP/Wide World Photos; C52 (l) George Hall/Woodfin Camp & Associates; C53 J. Aronovsky/Zuma Images/The Stock Market; C54 (tr) Courtesy of Scott Rowland; (bl) Dennis Oda/Tony Stone Images; C58-C59 (b) Lynn M. Stone/Bruce Coleman, Inc.; C59 (br) NASA; C60 Ann Duncan/Tom Stack & Associates; C63 (all) Bruce Coleman, Inc.; C66 Grant Heilman/Grant Heilman Photography; C68-C69 (b) Gary Irving/Panoramic Images; C68 (l) Barry L. Runk/Grant Heilman Photography; C69 (b) Barry L. Runk/Grant Heilman Photography; C70-C71 (b) Larry Lefever/Grant Heilman Photography; C72 Andy Sacks/Tony Stone Images; C74 USDA - Soil Conservation Service; C74-C75 (b) Dr. E.R. Degginger/Color-Pic; C75 (t) James D. Nations/D. Donne Bryant; (tli) Gunter Ziseler/Peter Arnold, Inc.; (tri) S.A.M./Wolfgang Kaehler Photography; (bli) Walter H. Hodge/Peter Arnold, Inc.; (bri) Jim Steinberg/Photo Researchers; C76 (t) Thomas Hovland from Grant Heilman Photography; (b) B.W. Hoffmann/AGStock USA; C78 (b) Randall B. Henne/Dembinsky Photo Associates; (l) Russ Munn/AgStock USA; C79 Bruce Hands/Tony Stone Images; C80 (tr) Courtesy of Diana Wall, Colorado State University; (bl) Oliver Mickes/Ottawa/Photo Researchers; (br) Kirby, Richar OSF/Earth Scenes; C86 Bob Daemmrich/Bob Daemmrich Photography, Inc.; C88 (l) Peter Correz/Tony Stone Images; (c) Mark E. Gibson; C88-C89(b) Bill Lea/Dembinsky Photo Associates; C90 (t) Chris Rogers/Rainbow/PNI; (c) Yoav Levy/Phototake/PNI; C91 Rob Badger Photography; C92 (t) Christie's Images, London/Superstock; (bc) Jeff Greenberg / Photo Researchers; (tr) Mary Ann Kulla/ The Stock Market; C93 (bl) Alan L. Detrick / Photo Researchers; (bc) Archive Photos; (br) David Barnes / The Stock Market; C94-C95 Jeff Greenberg/Visuals Unlimited; C95 (tl) Wolfgang Fischer/Peter Arnold, Inc.; (r) Wolfgang Fischer/Peter Arnold, Inc.; (c) Craig Hammell/Photo Researchers; C96 (t) Barbara Gerlach/Dembinsky Photo Associates; (b) Brownie Harris/The Stock Market; C97 Chris Rogers/The Stock Market; C98 Michael A. Keller/The Stock Market; C100-C101 (b) Ray Pfortner/Peter Arnold, Inc.; C103 William E. Ferguson; C104-C105 Blaine Harrington III/The Stock Market; C106 James King-Holmes/Science Photo Library/Photo Researchers; C107 (b) Wellman Fibers Industry; (r) Gabe Palmer/The Stock Market; C108 (tr) Susan Sterner/HRW; (bl) Kristin Finnegan/Tony Stone Images.

Unit D Other - D1(bg) Pal Hermansen/Tony Stone Images; (i) Earth Imaging/Tony Stone Images; D2-D3 (bg) Zefa Germany/The Stock Market; D3 (tr) Michael A. Keller/The Stock Market; (br) Steven Needham/Envision; D4 J. Shaw/Bruce Coleman, Inc.; D5 (b) NASA; D6 (l) Yu Somov; S. Korytnikov/Sovfoto/Eastfoto/PNI; (r) Christopher Arend/Alaska Stock Images/PNI; D7 Grant Heilman Photography; D8-D9 Dr. Eckart Pott/Bruce Coleman, Inc.; D10 N.R. Rowan/The Image Works; D12-D13 Mike Price/Bruce Coleman, Inc.; D13 David Job/Tony Stone Images; D14 John Beatty/Tony Stone Images; D17 (l) Grant Heilman Photography; (r) Darrell Gulin/Tony Stone Images; D20 NASA; D21 Ben Osborne/Tony Stone Images; D22 (r) Polytechnic State University; (b) NASA; D26-D27 Andrea Booher/Tony Stone Images; D27 NASA/Science Photo Library/Photo Researchers; D28 Joe Towers/The Stock Market; D30 Rich Iwasaki/Tony Stone Images; D32 (t) Peter Arnold; (c) Warren Faidley/International Stock Photography; (b) Stephen Simpson/FPG International; D33 E.R. Degginger/Color-Pic; D34 Ray Pfortner/Peter Arnold, Inc.; D37 (t) Ralph H. Wetmore, II/Tony Stone Images; (b) Joe McDonald/Earth Scenes; D38 (t) Tom Bean; (b) Adam Jones/Photo Researchers; D42 Warren Faidley/International Stock Photography; D44 (l) Superstock; (r) David Ducros/Science Photo Library/Photo Researchers; D45 (both) © 1998 AccuWeather; D45 New Scientist Magazine; D49 Dwayne Newton/PhotoEdit; D50 (tr) Courtesy June Bacon-Bercey; (bl) David R. Frazier/Photo Researchers; D54-D55 David Hardy/Science Photo Library/Photo Researchers; D55 (tc) European Space Agency/Science Photo Library/Photo Researchers; D60 (t) U.S. Geological Survey/Science Photo Library/Photo Researchers; D60 (b) NASA; D60, D61, D62 (bg) Jerry Schad/Photo Researchers; D61 (t) National Oceanic and Atmospheric Administration; D61 (b) David Crisp and the WFPC2 Science Team (Jet Propulsion Laboratory/California Institute of Technology); D62 (t), (b) NASA; D63 (t) Erich Karkoschka (University of Arizona Lunar & Planetary Lab) and NASA; (b) NASA; D64 J. Spurr/Bruce Coleman, Inc.; D65 Royer, Ronald/Science Photo Library/Photo Researchers; D66 Renee Lynn/Photo Researchers; D69 (l) Dr. E. R. Degginger/Color-Pic; (r) Dr. E. R. Degginger/Color-Pic; D72 (t) Joseph Nettis/Photo Researchers; (b) John Elk III/Bruce Coleman, Inc.; D74 NASA; D77 (all) Telegraph Colour Library/FPG International; D78-D79 Margaret Miller/Photo Researchers; D79 (t) Pekka Parviainen/Science Photo Library/Photo Researchers; (bg) George East/Science Photo Library/Photo Researchers; D80 Dr. Fred Espenak/Science Photo Library/Photo Researchers; D82 The Granger Collection, New York; D88 Merritt Vincent/PhotoEdit; D90 (l) Rob Talbot/ Tony Stone Images; (r) Stephen Graham/Dembinsky Photo Associates; D91 NASA; D92 (both) NASA.

Unit E Other - E1(bg) Steve Barnett/Liaison International; (i) StockFood America/Lieberman; E2-E3 Chris Noble/Tony Stone Images; E3 Kent & Donna Dannen; E6 John Michael/International Stock Photography; E7 (r) R. Van Nostrand/Photo Researchers; E8 (tr) Mike Timo/Tony Stone Images; E9 (t) Daniel J. Cox/Tony Stone Images; (b) Goknar/Vogue/Superstock; E10 (tr) John Michael/International Stock Photography; (bc) William Cornett/Image Excellence Photography; E11 (tr) Lee Foster/FPG International; (br) Paul Silverman/Fundamental Photographs; E12 (t) William Johnson/Stock, Boston; E12-E13 (b) Robert Finken/Photo Researchers; E14 S.J. Krasemann/Peter Arnold, Inc.; E16 (tr) Dr. E. R. Degginger/Color-Pic; E17 (l) Charles D. Winters/Photo Researchers; (br) Spencer Gran/PhotoEdit; E18-E19 Peter French/Pacific Stock; E20 J. Sebo/Zoo Atlanta; E25 (cr) Robert Pearcy/Animals Animals; E25 (l) Ron Kimball Photography; E26 (tr) Jim Harrison/Stock, Boston; E32 (tr) Corbis; (bl) Alfred Pasieka/Science Photo Library/Photo Researchers; E36-E37 (bg)Dr. Dennis Kunkel/Phototake; E38 Robert Ginn/PhotoEdit; E42 (tl), (tr) Dr. E. R. Degginger/Color-Pic; (b) Tom Pantages; (t) Tom Pantages; E44 Chip Clark; E46 (t) Tom Pantages; (c) Tom Pantages; E47 (bl) John Kaprielian/Tony Stone Images; E50 John Gaudio; E51 Michael Newman/Photo Edit; E52 (t) Los Alamos National Laboratory/Photo Researchers; (bl) US Army White Sands Missile Range.

Unit F - F1 (bg) Simon Fraser/Science Photo Library/Photo Researchers; (i) Nance Trueworthy/Liaison International; F2 (bg) April Riehm; (tr) M. W. Black/Bruce Coleman, Inc.; F6 (bl) Mary Kate Denny/PhotoEdit; (br) Gary A. Conner/PhotoEdit; F7 (b) Camerique, Inc./The Picture Cube; (tr) Stephen Saks/The Picture Cube; F8 (cr) Pat Field/Bruce Coleman, Inc.; (bl) Mark E. Gibson; F9 (l) Dr. E. R. Degginger/Color-Pic; (r)Ryan and Beyer/Allstock/PNI; F10 (b) John Running/Stock Boston; (cl) Joseph Nettis/Photo Researchers; F11 (b) Jeff Schultz/Alaska Stock Images, Stock Boston; F16 (br) Buck Ennis/Stock, Boston; (b) J.C. Carton/Bruce Coleman, Inc.; F24 (tl) Michael Holford Photographs; (br) Spencer Grant/PhotoEdit; F25 Marco Cristofori/ The Stock Market; F26 (tr) Corbis; (b) Shaun Egan/Tony Stone Images; F30-F31 (bg) Jerry Lodriguss/Photo Researchers; F31 (br) Picture PerfectUSA; F32 (b) James M. Mejuto/Photo Researchers; F34 (b) Mark E. Gibson; F36 (b) Bob Daemmrich/Stock, Boston; F36 (b) Myrleen Ferguson/PhotoEdit; F37 (b), (tl) Jan Butchofsky/Dave G. Houser; F39 (tr), (c) Richard Megna/Fundamental Photographs; F42 (b) Randy Duchaine/The Stock Market; F44 (tr) Tom Skrivan/The Stock Market; F45 (tr) David Woodfall/Tony Stone Images; F47 (b) Roy Morsh/The Stock Market; F48 (l) Ed Eckstein for The Franklin Institute Science Museum; (r) Peter Angelo Simon/The Stock Market; F48-F49 Paul Silverman/ Fundamental Photography; F54-F55 (bg) Superstock; (c) David Madison/Bruce Coleman, Inc.; F58 (cl) John Running/Stock, Boston; (b) David Young-Wolff/PhotoEdit; F59 H. Mark Weidman; F60 (t) D & I McDonald/The Picture Cube; F62 (b) Nasa/The Stock Market; (r) Richard Megna/Fundamental Photographs; F64 (b) Edith G. Haun/Stock, Boston; F70 (b) Amy C. Etra/PhotoEdit; F71 (b) Tony Freeman/PhotoEdit; F72 (b) Dave G. Houser; F74 Webb Chappell; F75 John Lei/Omni-Photo Communications; F76 (tr) NASA/Langley Research Center; (bl) Valder/Tormey/International Stock.

Health Handbook - R15 Palm Beach Post; R19 (tr) Andrew Speilman/Phototake; (c) Martha McBride/Unicorn; (br) Larry West/FPG International; R21 Superstock; R26 (c) Index Stock; R27 (t) Renee Lynn/ Tony Stone Images; (tr) David Young-Wolff/PhotoEdit.

All Other photographs by Harcourt photographers listed below, © Harcourt:
Weronica Ankarorn, Bartlett Digital Photographers, Victoria Bowen, Eric Camdem, Digital Imaging Group, Charles Hodges, Ken Karp, Ken Kinzie, Ed McDonald, Sheri O'Neal, Terry Sinclair.

Illustration Credits - Craig Austin A53; Graham Austin B56; John Butler A42; Rick Courtney A20, A51, B14, B55, C22, C40, C41, C62, C64; Mike Dammer A27, A61, B35, B61, C27, C55, C81, C109, D23, D51, D93, E35, E53, F27, F51, F77; Dennis Davidson D58; John Edwards D9, D17, D18, D31, D37, D39, D40, D59, D64, D68, D69, D70, D71, D76, D77, D78, D80, E17, D10, F45, F60; Wendy Griswold-Smith A37; Lisa Frasier F78; Geosystems C36, C42, C44, C103, D12, D46; Wayne Hovice B28; Tom Powers C8, C13, C20, C34, C50, C102; John Rice D16, B7, B21, B50; Ruttle D36, D7; Rosie Saunders A15; Shough C90, F22, F46, F71, F72.